6th BioASQ Workshop 2018

A Challenge on Large-Scale Biomedical Semantic Indexing and Question Answering

Brussels, Belgium
1 November 2018

ISBN: 978-1-5108-7420-6

EMNLP 2018

The 6th BioASQ Workshop
A challenge on large-scale biomedical semantic indexing and
question answering

Proceedings of the Workshop

November 1st, 2018
Brussels, Belgium

Order copies of this and other ACL proceedings from:

Association for Computational Linguistics (ACL)
209 N. Eighth Street
Stroudsburg, PA 18360
USA
Tel: +1-570-476-8006
Fax: +1-570-476-0860
acl@aclweb.org

Preface

The sixth BioASQ workshop on biomedical semantic indexing and question answering took place in Brussels, Belgium on November 1st, 2018 as part of the EMNLP 2018 conference and was hosted by the SQUARE Brussels meeting centre. The workshop was supported by the BioASQ project[1], which organizes the corresponding annual challenge. The goals of the workshop were to present the results of the sixth BioASQ challenge and further the interaction with the wider community of biomedical semantic indexing and question answering. The presenters represented research teams from different parts of the globe and with different viewpoints to the problem.This contributed to a very lively and interesting discussion among the participants of the workshop.

Ten papers were presented during the workshop. All were selected by peer review for presentation. This volume includes all ten papers. The first paper gives an overview of the challenge, including especially the datasets that were used throughout the challenges and the overall results achieved by the participants.

The remaining nine papers are those presented at the workshop. The first of these papers presents an analysis of the use of Semantic Role Labeling (SRL) tools in question-answering. The second paper present a system which uses deep learning and reinforcement learning approaches for the generation of ideal answers. Deep learning techniques for the document and snippet retrieval tasks is the object of discourse of the third workshop paper. The fourth paper presents a system for document and snippet retrieval, which makes use of semantic similarity patterns that are evaluated and measured by a convolutional neural network architecture. The system presented in the fifth paper uses a novel end-to-end model, which utilizes deep learning and attention mechanism to index MeSH terms to biomedical text. In the sixth paper, the authors move toward abstractive summarization, which attempts to distill and present the meaning of the original text in a more coherent way. They incorporate a sentence fusion approach, based on Integer Linear Programming. A named-entity based method for answering factoid and list questions, and an extractive summarization technique for building paragraphs are presented in the seventh paper of the workshop. The eighth paper studies the influence of enriching the training data by manually annotated variants of gold standard answers on the evaluation performance. The last paper focuses on a system for ideal answer generation, using ontology-based retrieval and a neural learning-to-rank approach, combined with extractive and abstractive summarization techniques.

We wish to thank all who participated to the success of this workshop, especially the authors, reviewers, speakers and participants.

Ioannis A. Kakadiaris, George Paliouras and Anastasia Krithara
November 2018

---

[1]www.bioasq.org

Organizers:

Ioannis A. Kakadiaris, University of Houston, USA
George Paliouras, NCSR "Demokritos", Greece and University of Houston, USA
Anastasia Krithara, NCSR "Demokritos", Greece

Program Committee:

Ion Androutsopoulos, Athens University of Economics and Business, Greece
Nicolas Baskiotis, Université Pierre et Marie Curie, France
Dimitris Galanis, National Technical University of Athens, Greece
Aris Kosmopoulos, NCSR "Demokritos", Greece
Sherri Matis-Mitchell, Consultant Text and Data Analytics at BioTraceIT, USA
Prodromos Malakasiotis, Athens University of Economics and Business, Greece
Jim Mork, National Library of Medicine, USA
Diego Molla, Macquarie University, Australia
Henning Müller, University of Applied Sciences, Switzerland
Claire Nedellec, INRA, France
Mariana Neves, University of Potsdam, Germany
Harris Papageorgiou, ILSP, Greece
Ioannis Partalas, Viseo group, France
John Prager, Thomas J. Watson Research Center, IBM, USA
Francisco J. Ribadas-Pena, University of Vigo, Spain
Hagit Shatkay, University of Delaware, USA
Grigoris Tsoumakas, Aristotle University of Thessaloniki, Greece

Invited Speaker:

Jin-Dong Kim, DBCLS, Japan

Table of Contents

Conference Program

9:00-9:10	**Welcome**
9:10-10:10	**Invited Talk: "Towards Explainable Question-Answering for Biomedical Applications"**
	Jin-Dong Kim, DBCLS, Japan
10:10-10:30	*Results of the 6th edition of BioASQ Challenge*
	Anastasios Nentidis, Anastasia Krithara, Kostas Bougiatiotis, Georgios Paliouras and Ioannis Kakadiaris
10:30-11:00	**Coffee break**
11:00-11:20	*Semantic role labeling tools for biomedical question answering: a study of selected tools on the BioASQ datasets*
	Fabian Eckert and Mariana Neves
11:20-11:40	*Macquarie University at BioASQ 6b: Deep learning and deep reinforcement learning for query-based summarisation*
	Diego Molla
11:40-12:00	*AUEB at BioASQ 6: Document and Snippet Retrieval*
	George Brokos, Polyvios Liosis, Ryan McDonald, Dimitris Pappas and Ion Androutsopoulos
12:00-12:20	*MindLab Neural Network Approach at BioASQ 6B*
	Andrés Rosso-Mateus, Fabio A. González and Manuel Montes-y-Gómez
12:20-14:00	**Lunch break**
14:00-14:20	*Ontology-Based Retrieval & Neural Approaches for BioASQ Ideal Answer Generation*
	Ashwin Naresh Kumar, Harini Kesavamoorthy, Madhura Das, Pramati Kalwad, Khyathi Chandu, Teruko Mitamura and Eric Nyberg
14:20-14:40	*Extraction Meets Abstraction: Ideal Answer Generation for Biomedical Questions*
	Yutong Li, Nicholas Gekakis, Qiuze Wu, Boyue Li, Khyathi Chandu and Eric Nyberg
14:40-15:00	*UNCC QA: Biomedical Question Answering system*
	Abhishek Bhandwaldar and Wlodek Zadrozn
15:00-15:20	*An Adaption of BIOASQ Question Answering dataset for Machine Reading systems by Manual Annotations of Answer Spans.*
	Sanjay Kamath, Brigitte Grau and Yue Ma
15:20-16:00	**Coffee break**
16:00-16:20	*AttentionMeSH: Simple, Effective and Interpretable Automatic MeSH Indexer*
	Qiao Jin, Bhuwan Dhingra, William Cohen and Xinghua Lu
16:20-16:30	**Award Announcements**
16:30-17:30	**Panel Discussion**

Results of the sixth edition of the BioASQ Challenge

Anastasios Nentidis[1,2]**, Anastasia Krithara**[1]**, Konstantinos Bougiatiotis**[1]**,
Georgios Paliouras**[1,3] **and Ioannis Kakadiaris**[3]
[1]National Center for Scientific Research "Demokritos", Athens, Greece
[2]Aristotle University of Thessaloniki, Thessaloniki, Greece
[3]University of Houston, Texas, USA

Abstract

This paper presents the results of the sixth edition of the BioASQ challenge. The BioASQ challenge aims at the promotion of systems and methodologies through the organization of a challenge on two tasks: semantic indexing and question answering. In total, 26 teams with more than 90 systems participated in this year's challenge. As in previous years, the best systems were able to outperform the strong baselines. This suggests that state-of-the-art systems are continuously improving, pushing the frontier of research.

1 Introduction

The aim of this paper is twofold. First, we aim to give an overview of the data issued during the BioASQ challenge in 2018. In addition, we aim to present the systems that participated in the challenge and evaluate their performance. To achieve these goals, we begin by giving a brief overview of the tasks, which took place from February to May 2018, and the challenge's data. Thereafter, we provide an overview of the systems that participated in the challenge. Detailed descriptions of some of the systems are given in workshop proceedings. The evaluation of the systems, which was carried out using state-of-the-art measures or manual assessment, is the last focal point of this paper, with remarks regarding the results of each task. The conclusions sum up this year's challenge.

2 Overview of the Tasks

The challenge comprised two tasks: (1) a large-scale semantic indexing task (Task 6a) and (2) a question answering task (Task 6b).

2.1 Large-scale semantic indexing - 6a

In Task 6a the goal is to classify documents from the PubMed digital library into concepts of the MeSH hierarchy. Here, new PubMed articles that are not yet annotated by MEDLINE indexers are collected and used as test sets for the evaluation of the participating systems. In contrast to previous years, articles from all journals were included in the test data sets of task 6a. As soon as the annotations are available from the MEDLINE indexers, the performance of each system is calculated using standard flat information retrieval measures, as well as, hierarchical ones. As in previous years, an on-line and large-scale scenario was provided, dividing the task into three independent batches of 5 weekly test sets each. Participants had 21 hours to provide their answers for each test set. Table 1 shows the number of articles in each test set of each batch of the challenge. 13,486,072 articles with 12.69 labels per article, on average, were provided as training data to the participants.

2.2 Biomedical semantic QA - 6b

The goal of Task 6b was to provide a large-scale question answering challenge where the systems had to cope with all stages of a question answering task for four types of biomedical questions: yes/no, factoid, list and summary questions (Balikas et al., 2013). As in previous years, the task comprised two phases: In phase A, BioASQ released 100 questions and participants were asked to respond with relevant elements from specific resources, including relevant MEDLINE articles, relevant snippets extracted from the articles, relevant concepts and relevant RDF triples. In phase B, the released questions were enhanced with relevant articles and snippets selected manually and the participants had to respond with *exact answers*, as well as with summaries in natural language (dubbed *ideal answers*). The task was split into five independent batches and the two phases for each batch were run with a time gap of 24 hours. In each phase, the participants received 100 ques-

Proceedings of the 2018 EMNLP Workshop BioASQ: Large-scale Biomedical Semantic Indexing and Question Answering, pages 1–10
Brussels, Belgium, November 1st, 2018. ©2018 Association for Computational Linguistics

Batch	Articles	Annotated Articles	Labels per Article
	7,240	6,639	11.67
	7,678	7,499	12.95
1	10,488	10,319	13.04
	6,225	6,073	12.32
	6,617	6,486	12.96
Total	38,248	37,016	12.65
	6,239	6,118	12.51
	7,152	6,803	12.75
2	7,113	6,575	12.75
	5,833	5,412	13.00
	7,379	6,606	12.65
Total	33,716	31,514	12.73
	6,469	5,768	12.58
	6,544	5,501	12.86
3	6,743	5,467	12.67
	8,487	5,615	12.70
	7,478	4,038	12.63
Total	35,721	26,389	12.69

Table 1: Statistics on test datasets for Task 6a.

tions and had 24 hours to submit their answers. Table 2 presents the statistics of the training and test data provided to the participants. The evaluation included five test batches.

Batch	Size	Documents	Snippets
Train	2,251	12.01	14.72
Test 1	100	4.06	6.02
Test 2	100	3.77	5.03
Test 3	100	3.97	4.80
Test 4	100	3.39	4.03
Test 5	100	3.94	5.07
Total	2,751	10.52	12.95

Table 2: Statistics on the training and test datasets of Task 6b. All the numbers for the documents and snippets refer to averages.

3 Overview of Participants

3.1 Task 6a

For this task, 11 teams participated and results from 42 different systems were submitted. In the following paragraphs we describe those systems for which a description was available, stressing their key characteristics. An overview of the systems and their approaches can be seen in Table 3.

The *"SNOKE"* system variants were developed

System	Approach
AttentionMeSH	RNN, w2v, attention scheme
AUTH	d2v, tf-idf, LLDA, SVM, ensembles
DeepMesh	d2v, tf-idf, MESHlabeler
Iria	bigrams, Luchene Index, k-NN, ensembles, UIMA ConceptMapper
SNOKE	search engine, UIMA ConceptMapper

Table 3: Systems and approaches for Task 6a. Systems for which no description was available at the time of writing are omitted.

as an UIMA (Tanenblatt et al., 2010) text and data mining workflow, combined with a heterogeneous database architecture, where different search strategies were adopted to automatically select probable MeSH terms. More specifically, the system is based on the ZB MED Knowledge Environment (Müller et al., 2017), while also utilizing the Snowball Stemmer (Agichtein and Gravano, 2000), to find matches between MeSH terms and words in the title and abstract of each target document.

The *"AttentionMeSH"* systems utilize deep learning and attention mechanisms which enable the models to associate textual evidence with annotations, thus providing interpretability at the word level. Firstly, they use a bidirectional gated recurrent unit to derive word representations with contextual information (Cho et al., 2014), to represent each document. At the same time, all MeSH terms are embedded using a technique that takes into account co-occuring MeSH terms in textually similar articles and finally an attention matrix (Mullenbach et al., 2018) is created based on the MeSH and word representations, leading to MeSH-specific article representations. This procedure allows the model to provide local interpretations of the predicted MeSH terms in relation to words of a specific article, raising the interesting subject of how explanations of an automatic MeSH indexer could further help human annotators in this task.

Other participating systems, including the *"DeepMeSH"* systems (Peng et al., 2016), the systems of the *"AUTH"* team (Papagiannopoulou et al., 2016) and the *"Iria"* systems (Ribadas-Pena et al., 2015) are based on the same techniques used by theirs systems for the previous version of the challenge which are summarized in Table 3 and described in the corresponding challenge overview (Nentidis et al., 2017). Similarly to the previous year, two systems developed by the National Library of Medicine (NLM) to assist the indexers in the annotation of MEDLINE articles, served as baselines for the semantic indexing task of the challenge. The Medical Text Indexer (MTI) (Mork et al., 2014) with some enchantments introduced in (Zavorin et al., 2016) and an extension of it, incorporating features of the winning system of the first BioASQ challenge (Tsoumakas et al., 2013).

3.2 Task 6b

The question answering task was tackled by 50 different systems, developed by 15 teams. In the first phase, which concerns the retrieval of information required to answer a question, 9 teams with 27 systems participated. In the second phase, where teams are requested to submit exact and ideal answers, 10 teams with 27 different systems participated. Four of the teams participated in both phases. An overview of the technologies employed by each team can be seen in Table 4.

The *"AUEB"* team that participated only in Phase A, used novel extensions of deep learning models for retrieving question-relevant documents and snippets. Firstly, they pre-trained word embeddings (Mikolov et al., 2013) on a very large collection of articles from MEDLINE/PubMed, while also implementing some pre-processing steps (stop-word removal, stemming (Krovetz, 1993), tokenization etc.). Then, for the document retrieval task they focused on the PACRR model of (Hui et al., 2017) and the DRMM model (Guo et al., 2016), while for snippets retrieval they utilized the ABCNN model (Yin et al., 2015). Alongside the extensions made on these models, they also deployed a clever post-processing scheme for snippet retrieval, as well as a model for initial document-retrieval based on BM25 (Robertson and Jones, 1976) for efficiency purposes.

Another approach based on deep learning methodologies for Phase A, focusing again on document and snippet retrieval, was proposed by the *"MindLaB"* team from the National University of Colombia. While for the document retrieval they use the BM25 model and ElasticSearch for efficiency, they train a Convolutional Neural Net-

Systems	Phase	Approach
Olelo	A, B	SRL toolkits (BioKIT, BioSmile, PathLSTM)
AUTH	A, B	MetaMap, LingPipe, Lucene Index, Stanford Parser
AUEB	A	BM25, w2v , DL (PACRR, DRMM, ABCNN)
USTB	A	Sequential Dependence Models, Ensembles
MindLab	A	ElasticSearch, BM25, POS-Tags, w2v, DL (CNN)
MQU	B	DL (LSTM), w2v, Regression models, Reinforcement Learning
Oaqa	B	Maximum Margin Relevance, w2v, Block Ordering, ILP
LabZhu	B	PubTator, Standford POS tool, ranking
UNCC	B	Metamap, Lexical Chaining
L2PS	B	SQUAD, DRQA (RNN, LSTM), GloVe

Table 4: Systems and approaches for Task 6b. Systems for which no information was available at the time of writing are omitted.

work (CNN) for snippet retrieval. As in the previous approach, they utilized a very large collection of PubMed Articles to train the CNN with similarity matrices of question-answer pairs. More specifically, they deploy similar pre-processing steps (tokenization, lowercasing, skip-gram embeddings (Moen and Ananiadou, 2013)) for the question and the document texts, however they also apply Part of Speech tagging to extract syntactical information regarding the terms. Based on the idea that not all terms are equally informative (Dong et al., 2015), they deploy a salience weighting scheme focusing on verbs, nouns and adjectives. Another interesting extension is the way final rankings of the snippets are generated based on a pseudo-relevance-feedback re-ranking step (Riezler et al., 2007).

In Phase B, the Macquarie University (*"MQU"*) team focused on ideal answers and explored ideas of reinforcement learning on deep learning models. Extending their previous work (Molla, 2017), they implemented different models under a regression setting for finding similar sentences to a question, based on the corresponding word2vec embeddings of the question-sentence pairs. They also experimented with different ways of utilizing these embeddings, notably using a bidirectional Recurrent Neural Networks with LSTM cells (Hochreiter and Schmidhuber, 1997) to equip the model with knowledge regarding the sentence position. Moreover, they also run interesting experiments using reinforcement learning towards the ROUGE score of the ideal answers, based on their previous work (Mollá-Aliod, 2017), but the results did not advocate for the use of such models.

The Carnegie Mellon University team ("*OAQA*"), focused also on ideal answer generation, building upon previous versions of the "*OAQA*" system (Chandu et al., 2017). They experimented with ways to improve the generated answer by extracting the most relevant non-redundant sentences from multiple documents and then re-ordering and fusing them to make the resulting text more human-readable and coherent. To this end, they tried different ordering algorithms for sentences and also made various improvements in different stages of the candidate sentences expansion, fusion and filtering procedure that was already used by their model. Among the notable additions is the use of an Integer Linear Program (ILP) module that is capable of fusing repeated content and simplifying complicated sentences, thus improving human readability.

Another system deployed by the same team focuses on answer generation using a knowledge graph and a neural learning-to-rank approach, combined with different summarization techniques. One of the novelties introduced is the creation of an ontology-based retrieval module for relevant snippets, through the relation extraction between biomedical entities found in the abstracts' texts (Abacha and Zweigenbaum, 2015). Also, different learning-to-rank approaches were explored (Qin et al., 2010; Cao et al., 2006, 2007) alongside both extractive (Allahyari et al., 2017) and abstractive (See et al., 2017) summarization techniques for the ideal answers generation.

An interesting approach comes from the "*L2PS*" team where they use an open-domain model (Chen et al., 2017), pre-trained on the SQUAD (Rajpurkar et al., 2016) dataset, and fine-tuned to the biomedical domain. An interesting difference with other deep learning approaches is the fact that the GloVe embeddings (Pennington et al., 2014) were the best amongst the ones tried. Moreover, they raise interesting questions regarding the effects of non-normalized answers (synonyms, abbreviations, multi-word answers) in the evaluation of different systems.

The "*UNCC*" team participated in Phase B, deploying lexical chaining techniques (Reeve et al., 2006) for sentence similarity and ranking to extract summaries from related snippets and efficiently fuse them in an ideal answer. They take advantage of the MetaMap tool (Aronson and Lang, 2010) for biomedical entity recognition and they also present a way to extend their methodology to factoid/list question answering in Phase A as well.

"*Olelo*" is one of the approaches that tackles both phases of the question answering task. More specifically, in Phase A Semantic Role Labeling (SRL) approaches for QA systems were utilized. These focus on the automatic extraction of predicate-argument structures (PAS) from both questions and document text, aimed at finding semantically related PAS between associated pairs. For Phase B, the system is built on top of the SAP HANA database and uses various NLP components, such as question processing, document and passage retrieval, answer processing and multi-document summarization based on previous approaches (Schulze et al., 2016) to develop a comprehensive system that retrieves relevant information and provides both exact and ideal answers for biomedical questions.

Other systems, including the "*USTB*" (Jin et al., 2017) and the "*LabZhu*" (Peng et al., 2015) systems employed the same techniques used by their systems for the previous version of the challenge, as summarized in Table 4 and described in the previous challenge overview (Nentidis et al., 2017). In this challenge too, the open source OAQA system proposed by (Yang et al., 2016) served as baseline for phase B. The system which achieved among the highest performances in previous versions of the challenge remains a strong baseline for the exact answer generation task. The system is developed based on the UIMA framework. ClearNLP is employed for question and snippet parsing. MetaMap, TmTool (Wei et al.,

System	Batch 1		Batch 2		Batch 3	
	MiF	LCA-F	MiF	LCA-F	MiF	LCA-F
AttentionMeSH	-	-	12.75	13	10	12.875
AttentionMeSH2	-	-	13.25	13.5	9.125	11.625
AttentionMeSH3	-	-	11.875	12	8.625	10.625
AttentionMeSH4	-	-	10.625	10.75	7.375	11.375
AttentionMeSH5	-	-	9.875	11	7.25	11
DeepMeSH1	3.75	4.75	4	5	9.75	10.75
DeepMeSH2	1.875	1.875	2	2	7.25	7.5
DeepMeSH3	2.625	2.625	3	3	5.75	6
DeepMeSH4	**1**	**1**	**1**	**1**	7.25	7.75
Default MTI	4.875	3.75	5	3.75	10.5	5.25
iria-1	9.75	9.75	13	13	15.75	15.75
iria-2	-	-	-	-	18.75	18.75
MeSHmallow-1	-	-	-	-	24	24
MeSHmallow-2	-	-	-	-	24	24
MeSHmallow-3	-	-	-	-	24	24
MTI First Line Index	6	6	8.25	7.5	13.75	9.75
Semantic NoSQL KE 1	-	-	16.25	16	-	-
Semantic NoSQL KE 2	-	-	15.75	17	-	-
Semantic NoSQL KE 3	-	-	19.5	20	-	-
Semantic NoSQL KE 4	-	-	17.5	18	-	-
Semantic NoSQL KE 5	-	-	18.5	19	-	-
UMass Amherst T2T	-	-	-	-	19.25	19.25
xgx	8.5	8.5	5.75	6	5.25	4
xgx0	-	-	8.5	7	**3.25**	**2**
xgx1	-	-	-	-	4.5	2.375
xgx2	-	-	-	-	3.5	3.875
xgx3	-	-	-	-	4.75	4.25

Table 5: Average system ranks across the batches of the Task 6a. A hyphenation symbol (-) is used whenever the system participated in fewer than 4 tests in the batch. Systems with fewer than 4 participations in all batches are omitted.

2016), C-Value and LingPipe (Baldwin and Carpenter, 2003) are used for concept identification and UMLS Terminology Services (UTS) for concept retrieval. The final steps include identification of concept, document and snippet relevance, based on classifier components and scoring, ranking and reranking techniques.

4 Results

4.1 Task 6a

Each of the three batches of Task 6a were evaluated independently. The classification performance of the systems were measured using flat and hierarchical evaluation measures (Balikas et al., 2013). The micro F-measure (MiF) and the Lowest Common Ancestor F-measure (LCA-F) were used to choose the winners for each batch (Kosmopoulos et al., 2013).

According to (Demsar, 2006) the appropriate way to compare multiple classification systems over multiple datasets is based on their average rank across all the datasets. On each dataset the system with the best performance gets rank 1.0, the second best rank 2.0 and so on. In case two or more systems tie, they all receive the average rank. Table 5 presents the average rank (according to MiF and LCA-F) of each system over all the test sets for the corresponding batches. Note, that the average ranks are calculated for the 4 best results of each system in the batch according to the rules of the challenge.

The results in Task 6a show that in all test batches and for both flat and hierarchical measures, some systems outperform the strong baselines. The *"DeepMeSH"* systems achieve the best performance in the first two batches, outperformed only by *"xgx"* systems in the third batch. More detailed results can be found in the online results page[1]. Comparison of these results with corresponding system results from previous years reveals the improvement of both the baseline and the top performing systems through the years of the competition as shown in Figure 1.

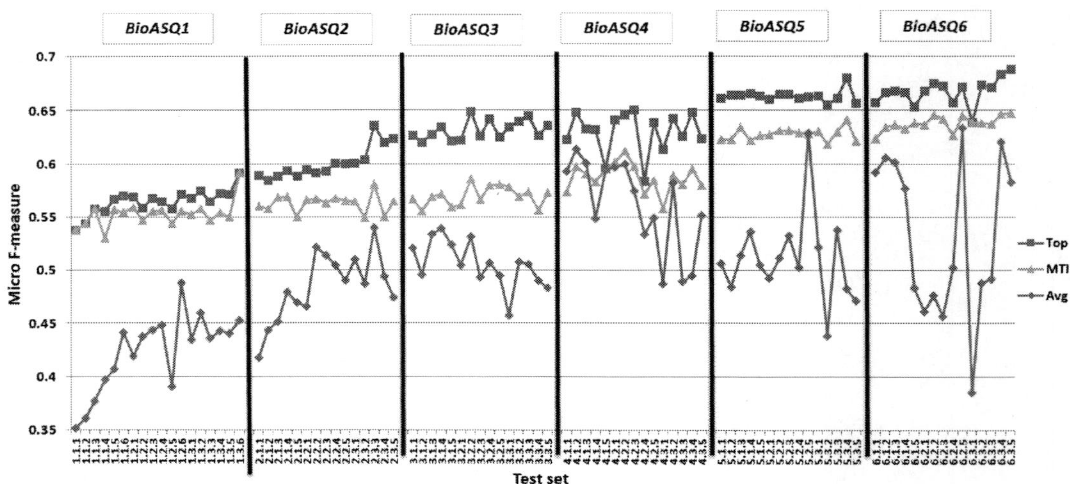

Figure 1: The micro f-measure achieved by systems across different years of the BioASQ challenge. For each test set the micro F-measure is presented for the best performing system (Top) and the MTI, as well as the average micro f-measure of all the participating systems (Avg).

System	Mean Precision	Mean Recall	Mean F-measure	MAP	GMAP
aueb-nlp-5	**0.2551**	**0.3412**	**0.2744**	**0.2314**	**0.0068**
MindLab QA Reloaded	0.1614	0.2657	0.1877	0.1344	0.0014
aueb-nlp-1	0.1384	0.2288	0.1563	0.1331	0.0046
aueb-nlp-3	0.1341	0.2263	0.1526	0.1294	0.0038
aueb-nlp-4	0.1325	0.2252	0.1519	0.1293	0.0038
aueb-nlp-2	0.1308	0.2204	0.1494	0.1262	0.0034
MindLab QA System	0.1542	0.2754	0.1833	0.1156	0.0023
MindLab Red Lions++	0.1406	0.2346	0.1636	0.1150	0.0013
MindLab QA System ++	0.1325	0.2252	0.1559	0.1148	0.0001
testtext	0.1802	0.2331	0.1831	0.1124	0.0035

Table 6: Results for snippet retrieval in batch 3 of phase A of Task 6b. Only the top-10 systems are presented.

4.2 Task 6b

Phase A: For phase A and for each of the four types of annotations: documents, concepts, snippets and RDF triples, we rank the systems according to the Mean Average Precision (MAP) measure. The final ranking for each batch is calculated as the average of the individual rankings in the different categories. In Tables 6 and 7 some indicative results from batch 3 are presented. Full results are available in the online results page of Task 6b, phase A[2]. These results are preliminary. The final results for Task 6b, phase A will be available after the manual assessment of the system responses.

Phase B: In phase B of Task 6b the systems were asked to produce exact and ideal answers. For ideal answers, the systems will eventually be ranked according to manual evaluation by the BioASQ experts (Balikas et al., 2013). Regarding

[1]http://participants-area.bioasq.org/results/6a/

[2]http://participants-area.bioasq.org/results/6b/phaseA/

System	Mean Precision	Mean Recall	Mean F-measure	MAP	GMAP
ustb_prir2	0.1660	**0.5674**	0.2186	**0.1281**	**0.0113**
ustb_prir3	**0.2007**	0.5609	**0.2496**	0.1259	0.0106
testtext	**0.2007**	0.5609	**0.2496**	0.1254	0.0106
ustb_prir4	0.1620	0.5601	0.2136	0.1253	0.0105
ustb_prir1	0.1700	0.5559	0.2203	0.1217	0.0100
aueb-nlp-2	0.1877	0.5352	0.2345	0.1147	0.0108
aueb-nlp-4	0.1877	0.5399	0.2345	0.1137	0.0106
aueb-nlp-3	0.1877	0.5429	0.2350	0.1135	0.0109
aueb-nlp-1	0.1877	0.5399	0.2345	0.1122	0.0101
sdm/rerank	0.1810	0.5422	0.2301	0.1061	0.0087

Table 7: Results for document retrieval in batch 3 of phase A of Task 6b. Only the top-10 systems are presented.

System	Yes/No		Factoid			List		
	Acc.	F1	Str. Acc.	Len. Acc.	MRR	Prec.	Rec.	F1
Oaqa-5b	**0.6667**	**0.6592**	0.0606	0.2121	0.1313	0.0867	0.2722	0.1299
fa2	0.6296	0.3864	0.2121	**0.3030**	**0.2475**	0.2511	0.3889	0.2955
fa4	0.6296	0.3864	0.2121	**0.3030**	0.2434	0.2800	0.3889	0.3131
fa1	0.6296	0.3864	0.2121	0.2727	0.2374	0.1600	0.4333	0.2290
fa3	0.6296	0.3864	0.2121	0.2727	0.2283	0.1800	**0.4778**	0.2564
Lab Zhu ,FDU	0.6296	0.3864	0.0909	0.1212	0.1061	0.1657	0.2833	0.1663
MQ-1	0.6296	0.3864	-	-	-	-	-	-
MQ-2	0.6296	0.3864	-	-	-	-	-	-
MQ-3	0.6296	0.3864	-	-	-	-	-	-
MQ-4	0.6296	0.3864	-	-	-	-	-	-
MQ-5	0.6296	0.3864	-	-	-	-	-	-
fa5	0.6296	0.5559	0.2121	**0.3030**	0.2434	0.2800	0.3889	0.3131
Lab Zhu,FDU	0.6296	0.3864	0.2121	0.2424	0.2273	0.2944	0.3444	0.2934
LabZhu,FDU	0.6296	0.3864	**0.2424**	0.2424	0.2424	**0.4130**	0.3389	**0.3312**
BioASQ_Baseline	0.4815	0.475	0.0606	0.1212	0.0859	0.1774	0.3944	0.2236

Table 8: Results for batch 4 for exact answers in phase B of Task 6b.

exact answers[3], the systems were ranked according to accuracy, F1 score on prediction of yes answer, F1 on prediction of no and macro-averaged F1 score for the yes/no questions, mean reciprocal rank (MRR) for the factoids and mean F-measure for the list questions. Table 8 shows the results for exact answers for the fourth batch of Task 6b. The symbol (-) is used when systems don't provide exact answers for a particular type of question. The full results of phase B of Task 6b are available online[4]. These results are preliminary. The final results for Task 6b, phase B will be available after the manual assessment of the system responses.

The results presented in Table 8 show that evaluation of system performance in the yes/no questions using the macro averaged F1 measure this year is useful to identify systems that achieve good performance regardless of any dataset imbalance in the yes-no classes. In batch 4 for example, two systems outperformed the strong baseline based on previous versions of the OAQA system, which is not clear considering only the accuracy. Regarding factoid and list questions, the performance achieved by the systems indicates that there is even more room for improvement in these types of question.

---

[3]For summary questions, no exact answers are required

[4]http://participants-area.bioasq.org/results/6b/phaseB/

5   Conclusions

In this paper, an overview of the sixth BioASQ challenge is presented. The challenge consisted of two tasks: semantic indexing and question answering. Overall, as in previous years, the best systems were able to outperform the strong baselines provided by the organizers. This suggests that advances over the state of the art were achieved through the BioASQ challenge but also that the benchmark in itself is challenging. Moreover, a clear shift towards the use of systems that incorporate ideas based on deep learning models can be seen, with respect to previous years. Novel ideas have been tested and state-of-the-art deep learning methodologies have been adapted to biomedical question answering with great results. Consequently, we believe that the challenge is successfully pushing the research frontier in biomedical information systems. In future editions of the challenge, we aim to provide even more benchmark data derived from a community-driven acquisition process.

Acknowledgments

The sixth edition of BioASQ is supported by a conference grant from the NIH/NLM (number 1R13LM012214-01) and sponsored by the Atypon Systems inc. BioASQ is grateful to NLM for providing baselines for task 6a and the CMU team for providing the baselines for task 6b. Finally, we would also like to thank all teams for their participation.

References

Asma Ben Abacha and Pierre Zweigenbaum. 2015. Means: A medical question-answering system combining nlp techniques and semantic web technologies. *Information processing & management*, 51(5):570–594.

Eugene Agichtein and Luis Gravano. 2000. Snowball: Extracting relations from large plain-text collections. In *Proceedings of the Fifth ACM Conference on Digital Libraries*, DL '00, pages 85–94, New York, NY, USA. ACM.

Mehdi Allahyari, Seyedamin Pouriyeh, Mehdi Assefi, Saeid Safaei, Elizabeth D Trippe, Juan B Gutierrez, and Krys Kochut. 2017. Text summarization techniques: a brief survey. *arXiv preprint arXiv:1707.02268*.

Alan R. Aronson and Franois-Michel Lang. 2010. An overview of MetaMap: historical perspective and recent advances. *Journal of the American Medical Informatics Association*, 17:229–236.

Breck Baldwin and Bob Carpenter. 2003. Lingpipe. *Available from World Wide Web: http://alias-i.com/lingpipe*.

Georgios Balikas, Ioannis Partalas, Aris Kosmopoulos, Sergios Petridis, Prodromos Malakasiotis, Ioannis Pavlopoulos, Ion Androutsopoulos, Nicolas Baskiotis, Eric Gaussier, Thierry Artieres, and Patrick Gallinari. 2013. Evaluation framework specifications. Project deliverable D4.1, UPMC.

Yunbo Cao, Jun Xu, Tie-Yan Liu, Hang Li, Yalou Huang, and Hsiao-Wuen Hon. 2006. Adapting ranking svm to document retrieval. In *Proceedings of the 29th annual international ACM SIGIR conference on Research and development in information retrieval*, pages 186–193. ACM.

Zhe Cao, Tao Qin, Tie-Yan Liu, Ming-Feng Tsai, and Hang Li. 2007. Learning to rank: from pairwise approach to listwise approach. In *Proceedings of the 24th international conference on Machine learning*, pages 129–136. ACM.

Khyathi Chandu, Aakanksha Naik, Aditya Chandrasekar, Zi Yang, Niloy Gupta, and Eric Nyberg. 2017. Tackling biomedical text summarization: Oaqa at bioasq 5b. *BioNLP 2017*, pages 58–66.

Danqi Chen, Adam Fisch, Jason Weston, and Antoine Bordes. 2017. Reading wikipedia to answer open-domain questions. *arXiv preprint arXiv:1704.00051*.

Kyunghyun Cho, Bart van Merrienboer, Çaglar Gülçehre, Fethi Bougares, Holger Schwenk, and Yoshua Bengio. 2014. Learning phrase representations using RNN encoder-decoder for statistical machine translation. *CoRR*, abs/1406.1078.

Janez Demsar. 2006. Statistical comparisons of classifiers over multiple data sets. *Journal of Machine Learning Research*, 7:1–30.

Li Dong, Furu Wei, Ming Zhou, and Ke Xu. 2015. Question answering over freebase with multi-column convolutional neural networks. In *Proceedings of the 53rd Annual Meeting of the Association for Computational Linguistics and the 7th International Joint Conference on Natural Language Processing (Volume 1: Long Papers)*, volume 1, pages 260–269.

Jiafeng Guo, Yixing Fan, Qingyao Ai, and W Bruce Croft. 2016. A deep relevance matching model for ad-hoc retrieval. In *Proceedings of the 25th ACM International on Conference on Information and Knowledge Management*, pages 55–64. ACM.

Sepp Hochreiter and Jürgen Schmidhuber. 1997. Long short-term memory. *Neural computation*, 9(8):1735–1780.

Kai Hui, Andrew Yates, Klaus Berberich, and Gerard de Melo. 2017. Pacrr: A position-aware neural ir model for relevance matching. *arXiv preprint arXiv:1704.03940*.

Zan-Xia Jin, Bo-Wen Zhang, Fan Fang, Le-Le Zhang, and Xu-Cheng Yin. 2017. A multi-strategy query processing approach for biomedical question answering: Ustb_prir at bioasq 2017 task 5b. *BioNLP 2017*, pages 373–380.

Aris Kosmopoulos, Ioannis Partalas, Eric Gaussier, Georgios Paliouras, and Ion Androutsopoulos. 2013. Evaluation Measures for Hierarchical Classification: a unified view and novel approaches. *CoRR*, abs/1306.6802.

Robert Krovetz. 1993. Viewing morphology as an inference process. In *Proceedings of the 16th annual international ACM SIGIR conference on Research and development in information retrieval*, pages 191–202. ACM.

Tomas Mikolov, Ilya Sutskever, Kai Chen, Greg S Corrado, and Jeff Dean. 2013. Distributed representations of words and phrases and their compositionality. In *Advances in neural information processing systems*, pages 3111–3119.

SPFGH Moen and Tapio Salakoski2 Sophia Ananiadou. 2013. Distributional semantics resources for biomedical text processing. In *Proceedings of the 5th International Symposium on Languages in Biology and Medicine, Tokyo, Japan*, pages 39–43.

Diego Molla. 2017. Macquarie university at bioasq 5b query-based summarisation techniques for selecting the ideal answers. In *Proceedings BioNLP 2017*.

Diego Mollá-Aliod. 2017. Towards the use of deep reinforcement learning with global policy for query-based extractive summarisation. In *Proceedings of the Australasian Language Technology Association Workshop 2017*, pages 103–107.

James G. Mork, Dina Demner-Fushman, Susan C. Schmidt, and Alan R. Aronson. 2014. Recent enhancements to the nlm medical text indexer. In *Proceedings of Question Answering Lab at CLEF*.

James Mullenbach, Sarah Wiegreffe, Jon Duke, Jimeng Sun, and Jacob Eisenstein. 2018. Explainable prediction of medical codes from clinical text. *CoRR*, abs/1802.05695.

Bernd Müller, Christoph Poley, Jana Pössel, Alexandra Hagelstein, and Thomas Gübitz. 2017. Livivo – the vertical search engine for life sciences. *Datenbank-Spektrum*, 17(1):29–34.

Anastasios Nentidis, Konstantinos Bougiatiotis, Anastasia Krithara, Georgios Paliouras, and Ioannis Kakadiaris. 2017. Results of the fifth edition of the bioasq challenge. *BioNLP 2017*, pages 48–57.

E Papagiannopoulou, Y Papanikolaou, D Dimitriadis, S Lagopoulos, G Tsoumakas, M Laliotis, N Markantonatos, and I Vlahavas. 2016. Large-scale semantic indexing and question answering in biomedicine. *ACL 2016*, page 50.

Shengwen Peng, Ronghui You, Hongning Wang, Chengxiang Zhai, Hiroshi Mamitsuka, and Shanfeng Zhu. 2016. Deepmesh: deep semantic representation for improving large-scale mesh indexing. *Bioinformatics*, 32(12):i70–i79.

Shengwen Peng, Ronghui You, Zhikai Xie, Yanchun Zhang, and Shanfeng Zhu. 2015. The fudan participation in the 2015 bioasq challenge: Large-scale biomedical semantic indexing and question answering. In *CEUR Workshop Proceedings*, volume 1391. CEUR Workshop Proceedings.

Jeffrey Pennington, Richard Socher, and Christopher Manning. 2014. Glove: Global vectors for word representation. In *Proceedings of the 2014 conference on empirical methods in natural language processing (EMNLP)*, pages 1532–1543.

Tao Qin, Tie-Yan Liu, Jun Xu, and Hang Li. 2010. Letor: A benchmark collection for research on learning to rank for information retrieval. *Information Retrieval*, 13(4):346–374.

Pranav Rajpurkar, Jian Zhang, Konstantin Lopyrev, and Percy Liang. 2016. Squad: 100,000+ questions for machine comprehension of text. *arXiv preprint arXiv:1606.05250*.

Lawrence Reeve, Hyoil Han, and Ari D Brooks. 2006. Biochain: lexical chaining methods for biomedical text summarization. In *Proceedings of the 2006 ACM symposium on Applied computing*, pages 180–184. ACM.

Francisco J. Ribadas-Pena, Luis M. de Campos, Víctor Manuel Darriba Bilbao, and Alfonso E. Romero. 2015. Cole and UTAI at bioasq 2015: Experiments with similarity based descriptor assignment. In *Working Notes of CLEF 2015 - Conference and Labs of the Evaluation forum, Toulouse, France, September 8-11, 2015*.

Stefan Riezler, Alexander Vasserman, Ioannis Tsochantaridis, Vibhu Mittal, and Yi Liu. 2007. Statistical machine translation for query expansion in answer retrieval. In *Proceedings of the 45th Annual Meeting of the Association of Computational Linguistics*, pages 464–471.

Stephen E Robertson and K Sparck Jones. 1976. Relevance weighting of search terms. *Journal of the American Society for Information science*, 27(3):129–146.

Frederik Schulze, Ricarda Schüler, Tim Draeger, Daniel Dummer, Alexander Ernst, Pedro Flemming, Cindy Perscheid, and Mariana Neves. 2016. Hpi question answering system in bioasq 2016. In *Proceedings of the Fourth BioASQ workshop at the Conference of the Association for Computational Linguistics*, pages 38–44.

Abigail See, Peter J Liu, and Christopher D Manning. 2017. Get to the point: Summarization with pointer-generator networks. *arXiv preprint arXiv:1704.04368*.

Michael A Tanenblatt, Anni Coden, and Igor L Sominsky. 2010. The conceptmapper approach to named entity recognition. In *LREC*.

Grigorios Tsoumakas, Manos Laliotis, Nikos Markontanatos, and Ioannis Vlahavas. 2013. Large-Scale Semantic Indexing of Biomedical Publications. In *1st BioASQ Workshop: A challenge on large-scale biomedical semantic indexing and question answering*.

Chih-Hsuan Wei, Robert Leaman, and Zhiyong Lu. 2016. Beyond accuracy: creating interoperable and scalable text-mining web services. *Bioinformatics (Oxford, England)*, 32(12):1907–10.

Zi Yang, Yue Zhou, and Nyberg Eric. 2016. Learning to answer biomedical questions: Oaqa at bioasq 4b. *ACL 2016*, page 23.

Wenpeng Yin, Hinrich Schütze, Bing Xiang, and Bowen Zhou. 2015. Abcnn: Attention-based convolutional neural network for modeling sentence pairs. *arXiv preprint arXiv:1512.05193*.

Ilya Zavorin, James G Mork, and Dina Demner-Fushman. 2016. Using learning-to-rank to enhance nlm medical text indexer results. *ACL 2016*, page 8.

Semantic role labeling tools for biomedical question answering:
a study of selected tools on the BioASQ datasets

Fabian Eckert and Mariana Neves *

Hasso Plattner Institute at University of Potsdam,
August-Bebel-Strasse 88, Potsdam 14482, Germany
`faeckert@gmail.com` , `marianalaraneves@gmail.com`

Abstract

Question answering (QA) systems usually rely
on advanced natural language processing components to precisely understand the questions
and extract the answers. Semantic role labeling (SRL) is known to boost performance
for QA, but its use for biomedical texts has
not yet been fully studied. We analyzed
the performance of three SRL tools (BioKIT,
BIOSMILE and PathLSTM) on 1776 questions from the BioASQ challenge. We compared the systems regarding the coverage of
the questions and snippets, as well as based
on pre-defined criteria, such as easiness of installation, supported formats and usability. Finally, we integrated two of the tools in a simple QA system to further evaluate their performance over the official BioASQ test sets.

1 Introduction

Question answering (QA) is one of the most
complex applications of natural language processing (NLP). QA systems need to precisely understand questions, in order to infer which information is being requested, and usually include steps
such as question type and expected answer detection (Athenikos and Han, 2010; Neves and Leser,
2015). Likewise, the candidate documents or snippets that potentially contain the answers also need
to be analyzed to extract the requested answer.
Therefore, such systems usually rely on various
NLP components, such as named-entity recognition, part-of-speech tagging and semantic parsing
(Athenikos and Han, 2010).

Semantic role labeling (SRL) is one of the most
popular tools to support QA systems (Shen and
Lapata, 2007). It consists of automatically identifying predicates and their arguments, the so-called
predicate-argument structures (PAS). For instance,

for the question *"How many genes does the human hoxD cluster contain?"*, BioKIT (Dahlmeier
and Ng, 2010), an SRL tool for biomedicine, correctly identified the following PAS: the predicate
contains and two arguments (Arg0 - *the human
hoxD cluster* and Arg1 - *How many genes*).

SRL is known for its potential to boost QA
performance when extracting PAS from both the
question and the text (e.g., snippets of sentences).
Ideally, the same (or semantically related) PAS
should be found in both of them in order to effectively support QA applications (Shen and Lapata,
2007). Hence, a good coverage is an important requirement for a tool to be suitable for QA. For our
given example question, one of the answer snippets provided by the BioASQ challenge (Tsatsaronis et al., 2015) was *The human HOXD complex
contains nine genes HOXD1, HOXD3, HOXD4,
HOXD8, HOXD9, HOXD10, HOXD11, HOXD12
and HOXD13, which are clustered from [...]*. The
following PAS was detected by BioKIT: the predicate *contains* and the arguments Arg0 *the human
HOXD complex* and Arg1 *nine genes*. In this example, there is a perfect match between the predicates from the question and the snippet. Further,
the values for the argument Arg0 are similar and
could be considered as a match too. The answer
nine genes is indeed contained in Arg1, which also
matches the argument type of the question word of
the sentence. This example demonstrates how QA
systems can benefit from PASs that were automatically detected by an SRL tool. However, language
is more complex than reflected in this example.
Thus, besides performing SRL, further challenges
arise to integrate SRL and gain significant advantages in QA systems.

We are not aware of a comprehensive evaluation of available SRL tools on the BioASQ
dataset, which is the most comprehensive dataset
on biomedical QA (Tsatsaronis et al., 2015). We

---

* Current address: German Federal Institute for Risk
Assessment, Diedersdorfer Weg 1, Berlin 12277, Germany

11

Proceedings of the 2018 EMNLP Workshop BioASQ: Large-scale Biomedical Semantic Indexing and Question Answering, pages 11–21
Brussels, Belgium, November 1st, 2018. ©2018 Association for Computational Linguistics

investigated three SRL tools, two of which were specifically developed for the biomedical domain, namely, BioKIT (Dahlmeier and Ng, 2010) and BIOSMILE (Tsai et al., 2006), and one which is based on deep learning, i.e., PathLSTM (Roth and Lapata, 2016). The latter has neither been trained nor tuned to biomedicine but has recently achieved promising results on SRL. Our contribution in this work is three-fold: (i) we provide a comprehensive overview on SRL for biomedicine and QA; (ii) we perform a comparison of selected tools regarding pre-defined criteria based on hands-on experiments; and (iii) we evaluated the selected SRL tools on the BioASQ datasets regarding their PAS coverage and performance in a QA system.

In the next section we provide an overview on previous work on SRL for biomedical QA, followed by the methodology we defined for the selection, comparison and evaluation of the SRL tools. In section 4 we present our results and discussion, followed by the conclusions of this work.

2 Overview of SRL for biomedical question answering

SRL has been well researched in recent decades and various tools have been created in the meantime. In addition to the tools, researchers have proposed standards for PAS annotations, such as the PropBank annotation format with its corresponding corpus (Kingsbury and Palmer, 2003). This was the standard followed by most SRL tools, as mentioned in (Palmer et al., 2010). However, they also presented two other popular formats for the English language, with corresponding corpora: FrameNet (Baker et al., 1998) and VerbNet (Kipper-Schuler, 2005).

Various features have been explored when building SRL tools based on machine learning algorithms. In 2004, Hacioglu et al. published an SRL approach which was based on chunking (Hacioglu et al., 2004). They trained a Support Vector Machine (SVM) to perform a semantic chunk segmentation step and role labeling. In their publication, they presented a complex set of features, e.g., words and part-of-speech tags and named entities, and their annotations followed the PropBank format. In the same year, Xue et Palmer experimentally explored the influence of certain newly proposed features on SRL results (Xue and Palmer, 2004). They could achieve significant improvements, especially by including syntactic

frame features.

First efforts on neural-based SRL came a couple of years ago. In 2016, Roth et Lapata presented a novel SRL model that improved results of previous state of the art SRL tools for the open domain (Roth and Lapata, 2016). They utilized neural sequence modeling techniques and put special focus on improving the detection of nominal predicates. Their evaluation showed that the novel SRL model reached F_1-scores of 87.9% for in-domain data and 76.1% for out-of-domain data, thus improving the state of the art in both categories. The presented SRL tool is called PathLSTM and is publicly available with an up-to-date model.

Recently, Marcheggiani et al. published another neural model for dependency-based SRL (Marcheggiani et al., 2017). By applying a syntax-agnostic model, they could almost keep with the state of the art for in-domain data (F_1: 87.6%) and surpassed PathLSTM for out-of-domain data (F_1: 77.3%). Still last year, Do et al. discussed the role of implicit SRL and their approach to meet corresponding challenges (Do et al., 2017). Traditional SRL systems usually focused on explicit argument labels while implicit SRL aims at also finding the implicit ones. They used a recurrent neural semantic frame model for learning probability distributions over semantic argument sequences and could hereby improve the state of the art for detecting implicit semantic role labels.

2.1 Semantic Role Labeling on Biomedical Text Corpora

Since the last decade, numerous efforts have been made to apply SRL techniques to the biomedical domain. In 2004, Wattarujeekrit et al. published PASBio (Wattarujeekrit et al., 2004), a PropBank extension for the domain of molecular biology. The PASBio corpus contained PASs for a limited set of 30 predicate stems. The corpus was specifically designed to support the extraction of events in molecular biology.

A couple of years later, Chou et al. presented BioProp (Chou et al., 2006), a corpus with PASs for the biomedical domain. It is composed of approximately 500 articles from the GENiA corpus (Kim et al., 2003) which were annotated with PASs. The resulting corpus was used to train the biomedical SRL tool BIOSMILE. Initially, it supported finding PASs for the 30 predicates introduced by BioProp. Later, the tool was extended and trained to support a total of 82 predicates, which are listed on the tool's website.[1] For the publication on the BIOSMILE tool (Tsai et al., 2006), the authors compared the performance of the latter to their initial SRL tool, which was only trained on PropBank data from the newswire domain. Being tested on BioProp data, the initial SRL tool could only reach an overall F_1-score of 64.2% while BIOSMILE reached 87.1%.

Later on, in 2009, Barnickel et al. presented their biomedical SRL system called SENNA which was based on a neural network (Barnickel et al., 2009). They managed to outperform tools like BIOSMILE regarding processing time but could only reach a comparably small F_1-score of 54% in the biomedical domain. In the following year, Dahlmeier et al. published an article on domain adaptation for SRL in the biomedical domain and introduced their respective SRL tool: BioKIT (Dahlmeier and Ng, 2010). It was developed as an alternative to the lack of training data in the biomedical domain and to the expensiveness to create training datasets. The authors discuss why, in their opinion, the BioProp corpus alone was not sufficient to create a good SRL tool for biomedicine. One of the reasons they mentioned was that BioProp was limited to 30 predicates and that many PASs were not covered in the corpus. When training BioKIT, they relied on the 1,982 PASs from BioProp and another 90,000 PASs from PropBank. They evaluated six supervised domain adaptation algorithms and concluded that the InstPrune algorithm performed best and reached an F_1-score of 85.38%.

More recently, in 2015, Zhang et al. showed, that clinical SRL can also significantly benefit from integrating domain adaptation techniques (Zhang et al., 2015). They relied on PropBank and NomBank from the newswire domain and BioProp as their source domain datasets. For the target domain, they used a manually annotated clinical corpus. They compared and evaluated three state-of-the-art domain adaptation algorithms: instance pruning, transfer self-training and feature augmentation. Finally, in 2016, Zhang et al. published another study where they investigated how their domain adaptation techniques for the clinical domain would apply on top of different syntactic parsers and features (Zhang et al., 2016). The best

---

F_1-score they could reach on their clinical test data was 71.41%.

2.2 Semantic Role Labeling for Question Answering

In the past, different experiments and approaches for integrating SRL to QA systems have been elaborated. Some of them were partially related to the biomedical domain.

Shen et Lapata published an extensive study on the contribution of SRL to open-domain factoid QA (Shen and Lapata, 2007). Based on their experiments, they proposed a combination of syntactic and semantic annotations for the answer extraction part of QA. In general, they showed that QA can benefit from SRL but they also found much potential for preferable improvements. They pointed out that coverage is a key factor to achieve benefits from SRL for QA.

For their EPoCare QA system, Nio et al. created a role identification system for clinical QA using the PICO format (Niu et al., 2003). This role identification system showed similarities to the SRL task but was limited to a small set of task-specific roles and heavily based on the PICO format. Also in the biomedical domain, Shi et al. utilized SRL for their biomedical QA system for summary type questions (Shi et al., 2007). They basically used semantic role labels to measure semantic conformities in their sentence candidate ranking procedure. Therefore, they analyzed to which extend a sentence and the particular question contained matching PASs.

The biomolecular QA system by Lin et al. utilized BIOSMILE for detecting PAS both in the questions and in answer candidate sentences (Lin et al., 2008). The core of their QA system was a ranking module. Their results indicated that BIOSMILE in combination with named entity recognition is well suited for improving biomolecular QA systems. Nevertheless, biomolecular QA is a rather restricted domain with regard to biomedical QA. Finally, in the scope of the BioASQ challenge, our team experimented with the BioKIT tool for all four types of questions (factoid, list, yes/no and summary) (Neves et al., 2017). But the results we obtained with our simple approach were not very successful.

Type	No. questions	No. snippets
factoid	485	5,145
list	406	5,155
summary	392	4,069
yes/no	493	5,724
Total	1,776	20,093

Table 1: Statistics of the BioASQ training dataset for each question type.

3 Methods

In this section we describe the resources, tools and methodology that we used to select and analyze SRL tools for the biomedical QA.

3.1 BioASQ Dataset

We utilized the training dataset of 1,776 questions made available for the BioASQ challenge in 2017 (Tsatsaronis et al., 2015).[2] It combines test sets from the first four challenges. This dataset contains questions and the corresponding snippets of text which include the answer to the questions. The BioASQ dataset addresses four types of questions: factoid, list, summary and yes/no. Table 1 shows statistics on the number of questions and snippets. We considered all four question types in our analysis.

3.2 Criteria for the selection of SRL tools

Despite the many previous works on SRL for biomedicine (cf. Section 2), few tools are available for immediate use. Driven by time constraints, we decided to include the only two available SRL tools for biomedicine, i.e, BioKIT (Dahlmeier and Ng, 2010) and BioSMILE (Tsai et al., 2006), and one open domain SRL tool that has recently obtained state-of-the.art results, i.e., PathLSTM (Roth and Lapata, 2016). Due to the non-availability of an out-of-the-box working SRL tool based on their model, we did not include the tool developed by Marcheggiani et al. (Marcheggiani et al., 2017), even though it is freely available in GitHub. We give a short overview of the selected tools regarding their technical aspects:

BioKIT. It is available for the Linux operating systems and was mainly developed in Python and C.[3] BioKIT expects input as text files with line breaks as separators and outputs the SRL results in the CoNLL-09 format.[4] It can be built and com-

piled with Cmake if all required dependencies are previously installed.

BIOSMILE. It is available as a Web service and supports a REST API.[5] Requests via the API need to be in XML format and results are returned in the same format. As far as we know, the source code or binaries of BIOSMILE are not available.

PathLSTM. It is developed in Java, and the sources as well as a Java package are available in a GitHub repository.[6] It can be built via Apache Maven. Input and output formats are the standard ones for CoNNL.

3.3 Methodology for evaluation

We installed each tool (or accessed it via web service) and ran them on the questions and corresponding snippets of the BioASQ training dataset. The BIOSMILE API was rather slow and unstable, therefore, we did not manage to annotate the questions and snippets of the 4th year of the BioASQ challenge with BIOSMILE, which is part of the training dataset. Hence, we were only able to evaluate 1,308 questions and the corresponding 16,791 snippets for BIOSMILE. However, this should not significantly compromise the comparison between the tools, given that the BioASQ dataset appears to be very homogeneous.

We analyzed the tools with on three approaches: (a) an assessment based on pre-defined criteria (cf. Section 3.4); (b) an evaluation of the coverage by counting the numbers of questions and snippets for which PASs were found; and (c) performance of the tools in a simple QA system (cf. Section 3.5).

3.4 Evaluation criteria

We also analyzed the tools regarding some selected criteria:

Installation. It checks whether the tool could be easily installed or whether it required advanced skills for building, as well as whether we experienced any issues related to missing or outdated dependencies. This is important for a smooth integration into a QA system, given that the latter should not suffer from a lack of portability or maintainability after the integration.

---

[2] http://bioasq.org/

[3] http://nlp.comp.nus.edu.sg/software

[4] https://ufal.mff.cuni.cz/
conll2009-st/task-description.html

[5] http://bws.iis.sinica.edu.tw/BioC_
BIOSMILE/

[6] https://github.com/microth/PathLSTM

Support of standardized web API. It checks whether the tool offers a Web service and whether it could be accessed and used via API calls following standards, e.g., REST. This is important if no source or binaries are available to download. Additionally, this functionality constitutes a straightforward and easy way of integration without the use of own computational resources.

Input format. It specifies the supported standard input formats, e.g., XML, JSON or CoNLL. Standardized input formats can facilitate the integration process independent from the system's platform.

Output format. Similar to standardized input formats, it specifies the supported standard output formats, e.g., XML, JSON or CoNLL.

Parsing effort. It is our subjective rating on how easy it was to parse the content to and from the supported input and output formats.

Handling of special characters. It specifies whether the tool is able to handle special characters or if it runs into errors at presence of certain characters in the input text.

Speed. It assesses the tool's time performance for annotating questions and answer snippets. This should give an idea to which degree an integration of the respective SRL tool could slow down the whole QA system.

Robustness. It indicates how reliable the SRL tool behaves with regard to stability and accessibility. Issues with robustness might, for instance, be caused by the input or the unresponsiveness of a web service.

3.5 Integration of SRL tools into a QA system

We also evaluated the SRL tools in the context of a simple rule-based QA system. Our rules were designed to make use of SRL wherever possible, but we also included fall-back solutions for the case that no PAS were found (baseline system). We addressed three question types from the BioASQ challenge, namely yes/no, factoid and list questions. The rules and parameters were inspired and tuned by looking at the data from the first three years of the BioASQ challenge. Therefore, the evaluation of the SRL tools in our QA system was carried out only on the BioASQ dataset from the fourth year.

Table 2 gives an overview on different degrees of PAS matching in the rules for each question type. In general, the higher the level to which an answer snippet matches to a question, the higher is its relevance for the answer. More details on the rules that we defined for each question type are presented below.

Yes/No questions. In a first step, weights follow the matching schema in Table 2. In a second step, for each matching answer snippet with a relevance weight, we determined whether the answer is *yes* or *no* by analyzing the presence of negation terms close to the predicate. If a predicate was directly prefixed by the terms *"not"* or *"doesn't"*, its vote for the overall answer was *no* and received an initial weight boost of 1. If no negation terms were found in the answer snippet, the answer for this snippet was *yes*. Finally, the overall decision on the answer was decided by calculating the balance of the weighted *yes* and *no* votes. In case of no matching at all, the default answer is *yes* (fall-back solution).

Factoid questions. We focused on PASs whose predicate was present in the question and one argument that matched a question word, such as *"which"*,*"where"*, *"when"*, *"who"* or *"how"*. We followed the priority level from Table 2 by checking the matching predicates, argument types and contents. Candidate answers in the list were ordered according to the matching level. If there were no matching predicates between the answer snippets and the question, the list of answer candidates remained empty (no fall-back solution).

List questions. For list questions, and similar to factoid questions, we implemented a priority queue to detect arguments that probably contain the answer. The major difference between factoid and list questions is that list questions do not simply require a simple fact but an enumeration of facts that are relevant for the answer. This is taken into account by putting special attention on the recognition of symbols or words that usually indicate the presence of an enumeration inside the answer snippets. Therefore, we split the text of the arguments by commas and semicolons, as well as by the symbol '&' or the token *and*. Finally, if no predicate or PAS matches was found in any answer snippet, the system searched for any enumerations it could find (fall-back solution).

Priority level by PAS matching degree	Conditions per question type		
	yes/no	factoid	list
1	-	no matching predicate	
2	matching predicate		
3	level 2 + matching argument type		
4	level 3 + matching argument content		
Boosting Factors at each level	-	Argument type match with question word	
		Presence of enumerators	

Table 2: Overview on the PAS matching levels and corresponding weights for the various question types.

Criteria	BioKIT	BIOSMILE	PathLSTM
Installation	very hard	-	hard
Web API	no	yes	no
Input format	text	XML	text
Output format	CoNLL	XML	CoNLL
Parsing effort	high	normal	high
Spec. charact.	bad	good	bad
Speed	fast	variable	very fast
Robustness	stable	unstable	stable

Table 3: Comparison of the three SRL tools regarding the selected criteria.

4 Results and Discussion

In this section we provide an assessment of the tools regarding the pre-defined criteria, the PAS coverage and the QA integration. For all approaches, we provide a comprehensive discussion based on our hands-on experiments with the tools.

4.1 Evaluation by criteria

We present an evaluation of the three selected SRL tools on the previously defined criteria (cf Table 3) and provide a detailed discussion on our impressions for each tool.

BioKIT. It does not support a binary, executable package nor a Web service and, therefore, it needed to be built on the Linux operating system. It was admittedly very hard to build and compile BioKIT, given that it is mainly written in Python and C but also depends on other packages and languages, such as Fortran. Many dependencies were outdated or missing and had to be searched in the Web. Therefore, simply following the installation instructions was not sufficient as some of the dependencies were themselves hard to build. Further, parsing the CoNLL format was more challenging than parsing XML or JSON into an object-oriented representation because the PASs had to be extracted by dynamically matching row and column indexes of the presented predicates and arguments. Additionally, BioKIT failed at handling special characters which led to annoying runtime errors. Usually, BioKIT's preprocessing pipeline was meant to eliminate problematic characters but some symbols (e.g., "æ", "ö" or "ò") were not handled by the system. As a result, BioKIT crashed with an error after processing thousands of sentences without returning any result when it ran into a special character. We collected a set of almost 20 of such characters that we eliminated in an own script-based preprocessing step. Depending on the length of the question or snippet, the processing of one question or snippet took at least 600 milliseconds or few seconds. This could be rated as a fast performance, but only when labeling many questions at once. If BioKIT was just used to process a single question, its runtime exceeded one minute, given the necessary time to load models into memory. In spite of the problem with special characters, we found BioKIT to be reliable and stable.

BIOSMILE. It is not available to download in any way (source code, binaries or executables). Therefore, we accessed it via a Web service with the REST API. The input and output were both formatted as XML, which facilitated parsing with standard XML parsing libraries. Further, we experienced no problems regarding special characters. However, with regard to the processing speed, the web service was rather unstable. In rare cases, it was possible to annotate a sentence in about a second but there were many problems regarding the robustness of the service. Frequently, it was not possible to send more than five requests in a row without waiting several minutes in between, otherwise the Web service became unresponsive for a long time. At some point, the service became so slow and had so many down times that we did not manage to annotate the data of the 4th year of the BioASQ challenge.

PathLSTM. Installing PathLSTM was not as hard as BioKIT but there were still some time-consuming issues. The tool can be build via Maven but it was under development during the time that we were using it (as of June/2017). The code actually contained missing or wrong Maven dependencies and even a bug due to an outdated or not committed class. Hence, the installation process required certain research and code review to find out that an earlier git commit was working properly. Recently, the developers of PathLSTM published a more stable package but we did not check its feasibility yet. The input and output formats also followed the CoNLL format and, hence, were very similar to BioKIT. The CoNLL parser for BioKIT could be reused with small adaptations. PathLSTM had similar issues with special characters and could therefore reuse the preprocessing script that was created for BioKIT. When annotating many questions or answer snippets at once, PathLSTM could reach an annotation rate as low as 300 milliseconds per sentence. We considered it as being very fast, in comparison to the other tools. But if trying to annotate a single sentence, PathLSTM had the same issue as BioKIT and needed almost one and a half minute to load the models into memory.

4.2 Evaluation by predicate-argument structure coverage

This section compares the three SRL tools with regard to the usefulness and completeness of the detected PASs for the QA task. As pointed out in (Shen and Lapata, 2007), the PAS coverage of SRL systems is an important factor when trying to successfully integrate SRL into QA. Therefore, we compared the PAS coverage of each tool for questions and answer snippets from the BioASQ datasets. With special regard to the QA task, we analyzed the PAS matching coverage between questions and corresponding answer snippets. The PAS matching coverage is defined as the percentage of questions for which a PAS match could be found in any of the corresponding answer snippets.

PAS coverage for questions and answer snippets. Table 4 gives an overview on the PAS coverage reached by the respective tools for the various types of questions and for all answer snippets in general. Answer snippets are not presented by question types because they do not differ by question type. When comparing the coverage of different question types, the lowest coverage values were reached for summary questions, while the highest coverage values were reached for yes/no questions. Only BIOSMILE performed better on factoid and list questions than on yes/no questions. This is probably due to predicate stems such as *do* or *have* that are widely used in yes/no questions which not part of the predicates supported by BIOSMILE. BIOSMILE obtained the lowest coverage results of the three tools, especially when looking at the answer snippets, which only 15.2% of them had PAS annotations. For list questions, BIOSMILE reached a coverage of 65.1%, which was slightly higher than the coverage of BioKIT in the same category (61.8%). The main reason for the low coverage of BIOSMILE is most probably the limited set of 82 biomedical predicates.

BioKIT reached the maximum coverage for yes/no questions (99.8%). This could be due to the fact that *do* (727) and *be* (247) are the top predicate stems detected in the questions. In comparison to this, PathLSTM only labeled *do* 27 times as a predicate and never labeled *be*. Additionally, BioKIT also labeled auxiliary verbs as predicates, which appear very often in yes/no questions, and might explain its high coverage. Unfortunately, auxiliary verbs like *has* or *has been* are not known to have much semantic value. Hence, this high coverage might not be seen as an advantage for PathLSTM.

In general, PathLSTM obtained significantly higher coverage values than the other tools. In contrast to leaving out auxiliary verbs, the high coverage of PathLSTM can be explained by detecting about three times as many distinct predicates as BioKIT. In general, reaching a higher coverage might be good for QA, but by looking at some of the annotations, we found that PathLSTM labeled many nouns (as predicates) that did not even had in a verb form and most likely did not represent a predicate. For example, the most frequent predicates found by PathLSTM were nouns such as *disease* or *syndrome*, none of which are regularly used as predicates.

PAS matching coverage between questions and answer snippets We evaluated two levels of PAS matching coverage between questions and answer snippets. The first level, which is presented in Table 5, is the proportion of questions for which any answer snippet contained a PAS with the same predicate stem. The second level, which is presented in Table 6, requires that both predicate argument structures, from the question and from the particular answer snippet, share a similar argument type besides the predicate stem. The

Type	questions			answer snippets		
	BioKIT	BIOSMILE*	PathLSTM	BioKIT	BIOSMILE*	PathLSTM
factoid	61.4	49.5	96.3			
list	61.8	65.1	96.1	88.5	15.2	98.5
summary	39.5	19.3	90.3			
yes/no	99.8	42.7	96.1			

Table 4: PAS coverage (in %) for the various types of questions and for answer snippets. * BIOSMILE was only evaluated on data from the first three years of the BioASQ challenge.

Type	BioKIT	BIOSMILE*	PathLSTM
factoid	28.9	2.1	68.7
list	33.7	2.1	82.5
summary	16.3	0.7	70.9
yes/no	38.9	4.0	79.3

Table 5: Coverage of the questions (in %) for which a predicate match between the question and any of the related answer snippets was found. * BIOSMILE was only evaluated on data from the first three years of the BioASQ challenge.

Type	BioKIT	BIOSMILE*	PathLSTM
factoid	26.8	2.1	59.6
list	33.3	2.1	72.2
summary	13.8	0.7	60.5
yes/no	36.7	4.0	72.6

Table 6: Coverage of the questions (in %) for which a PAS match between the question and any of the related snippets was found. A PAS match was counted, if the predicate stem and any of the related argument types matched. * BIOSMILE was only evaluated on data from the first three years of the BioASQ challenge.

argument type and the predicate stem have to be related by the same predicate.

On both PAS matching levels, BIOSMILE reached very poor results, between 2% and 4% of PAS matching coverage. It appears that the same questions of Table 5 found by BIOSMILE also fulfill the requirements of Table 6, which might indicate that the found PAS matches are of a good quality. Nevertheless, the coverage is so low that BIOSMILE might only be considered in combination with other tools with a higher coverage in order to be efficiently used for biomedical QA. It would be pointless to exclusively rely on a tool that can only contribute to answering up to 4% of the questions.

In contrast, PathLSTM reached the highest PAS matching coverage values on both levels. Table 5 shows that PathLSTM obtained 82.5% PAS matching coverage for list questions with matching predicate stems. Further, Table 6 shows that

for 72.2% of the questions, a matching argument type was present. On the one hand, the high coverage of PathLSTM might lead to a high recall when implementing a QA system on top of the annotations. On the other hand, our previous analysis of PathLSTM's predicates (cf. above) showed that they might be of poor quality.

The PAS matching coverage for BioKIT were not as high as the results reached by PathLSTM but superior than those from BIOSMILE. BioKIT leaves some space for improvement regarding coverage by finding matches for about one third of the factoid, list and yes/no questions and less than one sixth for summary questions. It is striking that the differences between the PAS matching coverage values of both levels are not very large (below 2.5%). In contrast to PathLSTM, this might be an indicator that PAS matches found by BioKIT are actually of a good quality and semantically relevant, and not just simply include matching terms that are not even real predicates and hence have no related arguments. Finally, PathLSTM reached PAS matching coverage values which are in average more than twice as large as those from BioKIT, but the quality and usefulness of the PAS matches from PathLSTM are still dubious.

4.3 Evaluation on the rule-based QA system

We compared BioKIT and PathLSTM regarding their performance on the fourth year of the BioASQ challenge. This dataset is composed of five batches of 100 questions and we provide detailed results for each batch. The results were obtained by uploading JSON result files to the BioASQ Oracle evaluation system[7]. The BIOSMILE system was not further evaluated due to (i) its low PAS matching coverage (cf. Table 5), which were very unpromising, and (ii) the instability of the Web service, which did not allow us to obtain results for this test set.

---

[7] http://participants-area.bioasq.org/accounts/login/?next=/oracle/

Batch	BioKIT	PathLSTM	NOSRL
1		96.43	
2		90.63	
3		96.0	
4		90.48	
5		100.0	
Average		94.71	

Table 7: Evaluation of the accuracy (in %) for yes/no questions on the 5 batches of the fourth BioASQ challenge.

Batch	BioKIT	PathLSTM
1	8.97	1.79
2	1.62	0
3	5.13	0
4	6.72	1.61
5	1.36	0.76
Average	4.76	0.83

Table 8: Evaluation of the mean reciprocal rank (in %) for factoid questions on the 5 batches of the fourth BioASQ challenge.

To measure the impact of the SRL components added to the QA system, we included a baseline QA solution (NOSRL) which did not rely on SRL but simply only on the fall-back solutions (cf. Section 3.5). As we could not propose appropriate fall-back solutions for the factoid questions, we evaluated the NOSRL baseline only for yes/no and list questions.

Yes/no. Table 7 evaluates the accuracy of the SRL-based QA and the NOSRL systems. The latter simply answered *yes* to all questions. For all 5 batches of the fourth year of the BioASQ challenge, the SRL-based systems did not provide any answers different than *yes*. Therefore, all systems achieved the same accuracy values. Obviously, our rules failed to match any of the valid *no-*answers in the fourth year's dataset. Subsequently, we cannot provide insight with respect to the performance of the SRL tools.

Factoid. Table 8 compares the performance on factoid questions for the different SRL-based QA systems by means of the mean reciprocal rank measure (MRR). The results show that the BioKIT-based QA system performed much better than the PathLSTM-based version. For the second and third batch, the PathLSTM-based system did not find any correct answer.

List. Table 9 shows the mean average F-measure results for the five batches. In general, BioKIT performed better than the NOSRL system, while

Batch	BioKIT	PathLSTM	NOSRL
1	15.48	8.33	14.9
2	13.75	16.88	13.79
3	14.03	11.18	14.03
4	28.31	22.27	20.81
5	21.23	13.91	19.19
Average	18.56	14.51	16.54

Table 9: Evaluation of the mean average F-measure (in %) for list questions on the 5 batches of the fourth BioASQ challenge.

the PathLSTM-based system performed worse than the latter.

5 Conclusions and future work

Our experiments showed that BioKIT is the most suitable SRL tool for biomedical QA, and in a lesser degree, PathLSTM might also be considered. For both tools, different challenges might arise for their integration. While BioKIT still has a lack of coverage, PathLSTM probably detects too many PAS candidates and therefore performed poorly in our simple QA system. A first approach to increase the precision for PathLSTM would include filtering out noun predicates which do not have a verb stem. We would like to perform a more comprehensive evaluation of the quality of the PAS, given that we only carried out a small validation of a few of them. Recently, a new SRL tool (He et al., 2018) has been published and should also be considered in future experiments.

While BioSMILE is readily available, its web service is unstable and the coverage is extremely low. BioKIT is hard to install, but provides a good coverage of PAS which is suitable for the QA task. We assume that PathLSTM is too generic, as it is an open-domain SRL tool. It might therefore have trouble to compete with specialized biomedical SRL on data from the biomedical domain. Finally, even though the coverage from PathLSTM is high, an analysis of some of its PAS shows that many predicates have no semantic meaning and many correspond to nouns which do not behave as predicates in the corresponding sentences.

References

Sofia J. Athenikos and Hyoil Han. 2010. Biomedical question answering: A survey. *Computer Methods and Programs in Biomedicine*, 99(1):1 – 24.

Collin F Baker, Charles J Fillmore, and John B Lowe. 1998. The Berkeley FrameNet project. In *Proceedings of the 36th Annual Meeting of the Associ-*

ation for Computational Linguistics and 17th International Conference on Computational Linguistics-Volume 1, pages 86–90. Association for Computational Linguistics.

Thorsten Barnickel, Jason Weston, Ronan Collobert, Hans-Werner Mewes, and Volker Stümpflen. 2009. Large scale application of neural network based semantic role labeling for automated relation extraction from biomedical texts. *PLoS One*, 4(7):e6393.

Wen-Chi Chou, Richard Tzong-Han Tsai, Ying-Shan Su, Wei Ku, Ting-Yi Sung, and Wen-Lian Hsu. 2006. A semi-automatic method for annotating a biomedical proposition bank. In *Proceedings of the workshop on frontiers in linguistically annotated corpora 2006*, pages 5–12. Association for Computational Linguistics.

Daniel Dahlmeier and Hwee Tou Ng. 2010. Domain adaptation for semantic role labeling in the biomedical domain. *Bioinformatics*, 26(8):1098.

Quynh Ngoc Thi Do, Steven Bethard, and Marie-Francine Moens. 2017. Improving implicit semantic role labeling by predicting semantic frame arguments. *arXiv preprint arXiv:1704.02709*.

Kadri Hacioglu, Sameer Pradhan, Wayne H Ward, James H Martin, Daniel Jurafsky, et al. 2004. Semantic role labeling by tagging syntactic chunks. In *CoNLL*, pages 110–113.

Luheng He, Kenton Lee, Omer Levy, and Luke Zettlemoyer. 2018. Jointly predicting predicates and arguments in neural semantic role labeling. In *Proceedings of the 56th Annual Meeting of the Association for Computational Linguistics (Volume 2: Short Papers)*, pages 364–369. Association for Computational Linguistics.

Jin-Dong Kim, Tomoko Ohta, Yuka Tateisi, and Junichi Tsujii. 2003. GENIA corpus - a semantically annotated corpus for bio-textmining. *Bioinformatics*, 19(suppl 1):i180–i182.

Paul Kingsbury and Martha Palmer. 2003. Propbank: the next level of treebank. In *Proceedings of Treebanks and lexical Theories*, volume 3. Citeseer.

Karin Kipper-Schuler. 2005. *VerbNet: a broad-coverage, comprehensive verb lexicon*. Ph.D. thesis, Computer and Information Science Department, Universiy of Pennsylvania, Philadelphia, PA.

Ryan T. K. Lin, Justin Liang-Te Chiu, Hong-Jie Dai, Min-Yuh Day, Richard Tzong-Han Tsai, and Wen-Lian Hsu. 2008. Biological question answering with syntactic and semantic feature matching and an improved mean reciprocal ranking measurement. In *IRI*, pages 184–189. IEEE Systems, Man, and Cybernetics Society.

Diego Marcheggiani, Anton Frolov, and Ivan Titov. 2017. A simple and accurate syntax-agnostic neural model for dependency-based semantic role labeling.

In *Proceedings of the 21st Conference on Computational Natural Language Learning (CoNLL 2017)*, pages 411–420, Vancouver, Canada. Association for Computational Linguistics.

Mariana Neves, Fabian Eckert, Hendrik Folkerts, and Matthias Uflacker. 2017. Assessing the performance of olelo, a real-time biomedical question answering application. In *BioNLP 2017*, pages 342–350, Vancouver, Canada,. Association for Computational Linguistics.

Mariana Neves and Ulf Leser. 2015. Question answering for biology. *Methods*, 74:36 – 46. Text mining of biomedical literature.

Yun Niu, Graeme Hirst, Gregory McArthur, and Patricia Rodriguez-Gianolli. 2003. Answering clinical questions with role identification. In *Proceedings of the ACL 2003 Workshop on Natural Language Processing in Biomedicine - Volume 13*, BioMed '03, pages 73–80, Stroudsburg, PA, USA. Association for Computational Linguistics.

Martha Palmer, Daniel Gildea, and Nianwen Xue. 2010. Semantic role labeling. *Synthesis Lectures on Human Language Technologies*, 3(1):1–103.

Michael Roth and Mirella Lapata. 2016. Neural semantic role labeling with dependency path embeddings. In *Proceedings of the 54th Annual Meeting of the Association for Computational Linguistics (Volume 1: Long Papers)*, pages 1192–1202, Berlin, Germany. Association for Computational Linguistics.

Dan Shen and Mirella Lapata. 2007. Using semantic roles to improve question answering. In *EMNLP-CoNLL 2007, Proceedings of the 2007 Joint Conference on Empirical Methods in Natural Language Processing and Computational Natural Language Learning, June 28-30, 2007, Prague, Czech Republic*, pages 12–21. ACL.

Zhongmin Shi, Gabor Melli, Yang Wang, Yudong Liu, Baohua Gu, Mehdi M Kashani, Anoop Sarkar, and Fred Popowich. 2007. Question answering summarization of multiple biomedical documents. In *Advances in Artificial Intelligence*, pages 284–295. Springer.

Richard Tzong-Han Tsai, Wen-Chi Chou, Yu-Chun Lin, Cheng-Lung Sung, Wei Ku, Ying-Shan Su, Ting-Yi Sung, and Wen-Lian Hsu. 2006. BIOS-MILE: adapting semantic role labeling for biomedical verbs: an exponential model coupled with automatically generated template features. In *Proceedings of the HLT-NAACL BioNLP Workshop on Linking Natural Language and Biology*, pages 57–64. Association for Computational Linguistics.

George Tsatsaronis, Georgios Balikas, Prodromos Malakasiotis, Ioannis Partalas, Matthias Zschunke, Michael R Alvers, Dirk Weissenborn, Anastasia Krithara, Sergios Petridis, Dimitris Polychronopoulos, et al. 2015. An overview of the BIOASQ large-scale biomedical semantic indexing and question answering competition. *BMC bioinformatics*, 16(1):138.

Tuangthong Wattarujeekrit, Parantu K Shah, and Nigel Collier. 2004. PASBio: predicate-argument structures for event extraction in molecular biology. *BMC bioinformatics*, 5(1):155.

Nianwen Xue and Martha Palmer. 2004. Calibrating features for semantic role labeling. In *EMNLP*, pages 88–94.

Yaoyun Zhang, Min Jiang, Jingqi Wang, and Hua Xu. 2016. Semantic role labeling of clinical text: Comparing syntactic parsers and features. In *AMIA Annual Symposium Proceedings*, volume 2016, page 1283. American Medical Informatics Association.

Yaoyun Zhang, Buzhou Tang, Min Jiang, Jingqi Wang, and Hua Xu. 2015. Domain adaptation for semantic role labeling of clinical text. *J Am Med Inform Assoc*, 22(5):967–979. 26063745[pmid].

Macquarie University at BioASQ 6b: Deep learning and deep reinforcement learning for query-based multi-document summarisation

Diego Mollá
Macquarie University
Sydney, Australia
`diego.molla-aliod@mq.edu.au`

Abstract

This paper describes Macquarie University's contribution to the BioASQ Challenge (BioASQ 6b, Phase B). We focused on the extraction of the ideal answers, and the task was approached as an instance of query-based multi-document summarisation. In particular, this paper focuses on the experiments related to the deep learning and reinforcement learning approaches used in the submitted runs. The best run used a deep learning model under a regression-based framework. The deep learning architecture used features derived from the output of LSTM chains on word embeddings, plus features based on similarity with the query, and sentence position. The reinforcement learning approach was a proof-of-concept prototype that trained a global policy using REINFORCE. The global policy was implemented as a neural network that used $tf.idf$ features encoding the candidate sentence, question, and context.

1 Introduction

The BioASQ Challenge[1] consists of various tasks related to biomedical semantic indexing and question answering (Tsatsaronis et al., 2015). Our participation in BioASQ for 2018 focused on Task B Phase B, where our system attempted to find the ideal answer given a question and a collection of relevant snippets of text. We approached this task as an instance of query-based multi-document summarisation, where the ideal answer is the summary to produce.

The BioASQ challenge focuses on a restricted domain, namely biomedical literature. Nevertheless, the techniques developed for our system were domain-agnostic and can be applied to any domain, provided that the domain has enough training data and a specialised corpus large enough to train word embeddings.

[1] `http://www.bioasq.org/`

	Summary	Factoid	Yesno	List
n	6	2	2	3

Table 1: Value of n (the number of sentences returned as the ideal answer) for each question type.

We were interested in exploring the use of deep learning and reinforcement learning for this task. Thus, Section 2 explains our experiments using deep learning techniques. Section 3 details our experiments using reinforcement learning. Section 4 specifies the settings used in the experiments. Section 5 shows and discusses the results, and Section 6 concludes the paper.

2 Deep Learning

The deep learning experiments followed the general framework introduced by Mollá (2017a), which can be summarised as a regression approach that follows these three main steps:

1. Split the input text into candidate sentences.

2. Score each candidate sentence independently.

3. Return the n sentences that have the highest score.

In all of the experiments reported in this paper, the input text is the set of relevant snippets that are associated with the question. These snippets are pre-processed by splitting them into sentences. The sentences are then scored using the deep learning approaches described below. Then, after all candidate sentences are scored, the top n sentences are returned as the ideal answer. The value of n is determined empirically and it depends on the question type as shown in Table 1. These are the same settings as in Mollá (2017a)'s framework.

Proceedings of the 2018 EMNLP Workshop BioASQ: Large-scale Biomedical Semantic Indexing and Question Answering, pages 22–29
Brussels, Belgium, November 1st, 2018. ©2018 Association for Computational Linguistics

The deep learning experiments predict the score of an input sentence by applying supervised regression techniques. Following Mollá (2017a)'s framework, the training data was annotated with the F1 ROUGE-SU4 score of each individual sentence using the ideal answer as the target. Also following Mollá (2017a)'s framework, the architecture of our system architecture was based on the following main stages:

1. Obtain the word embedding of every word in the input sentence and the question.

2. Given the word embeddings of the input sentence and the question, obtain the sentence and question embeddings.

3. Feed the sentence embeddings and the similarity between the sentence and question embeddings to a fully connected layer and final linear combination. In this stage, as an extension to Mollá (2017a)'s approach, we also incorporated information about the sentence position.

The word embeddings were obtained by pretraining word2vec (Mikolov et al., 2013) on a collection of PubMed documents made available by the organisers of BioASQ. Given a sentence (or question) i with n_i words and vectors of word embeddings m_1 to m_{n_i}, we ran experiments using the following two alternative approaches to obtain the sentence (or question) vector of embeddings s_i:

NNR Mean. Compute the mean of the word embeddings:

$$s_i = \frac{1}{n_i} \sum_{j=1}^{n_i} m_j$$

NNR LSTM. Feed the sequence of word embeddings to bidirectional Recurrent Neural Networks with LSTM cells. We used TensorFlow's LSTM implementation, which is reportedly based on Hochreiter et al. (1997)'s seminal work. More explicitly, the embedding s_i of sentence i is the concatenation of the output of the last cell in the forward chain $(\overrightarrow{h}_{n_i})$ and the first cell in the backward chain $(\overleftarrow{h}_1)$:

$$s_i = \left[\overrightarrow{h}_{n_i}; \overleftarrow{h}_1 \right]$$

Cell at position t of the forward chain receives its input from the embedding of word at position t and cell at position $t - 1$:

$$\overrightarrow{c}_t = \overrightarrow{f} \odot \overrightarrow{c}_{t+1} + \overrightarrow{i} \odot \overrightarrow{z}$$
$$\overrightarrow{h}_t = \overrightarrow{o} \odot \tanh(\overrightarrow{c}_t)$$
$$\overrightarrow{i} = \sigma(\overrightarrow{W}_i \cdot m_t + \overrightarrow{U}_i \cdot \overrightarrow{h}_{t+1} + \overrightarrow{b}_i)$$
$$\overrightarrow{f} = \sigma(\overrightarrow{W}_f \cdot m_t + \overrightarrow{U}_f \cdot \overrightarrow{h}_{t+1} + \overrightarrow{b}_f)$$
$$\overrightarrow{o} = \sigma(\overrightarrow{W}_o \cdot m_t + \overrightarrow{U}_o \cdot \overrightarrow{h}_{t+1} + \overrightarrow{b}_o)$$
$$\overrightarrow{z} = \tanh(\overrightarrow{W}_z \cdot m_t + \overrightarrow{U}_z \cdot \overrightarrow{h}_{t+1} + \overrightarrow{b}_z)$$

where σ is the logistic sigmoid function, $\odot$ is the element-wise multiplication of two vectors, and $\cdot$ is the dot product between a matrix and a vector.

Cell at position t in the backward chain receives its input from m_t and cell at position $t + 1$:

$$\overleftarrow{c}_t = \overleftarrow{f} \odot \overleftarrow{c}_{t+1} + \overleftarrow{i} \odot \overleftarrow{z}$$
$$\overleftarrow{h}_t = \overleftarrow{o} \odot \tanh(\overleftarrow{c}_t)$$
$$\overleftarrow{i} = \sigma(\overleftarrow{W}_i \cdot m_t + \overleftarrow{U}_i \cdot \overleftarrow{h}_{t+1} + \overleftarrow{b}_i)$$
$$\overleftarrow{f} = \sigma(\overleftarrow{W}_f \cdot m_t + \overleftarrow{U}_f \cdot \overleftarrow{h}_{t+1} + \overleftarrow{b}_f)$$
$$\overleftarrow{o} = \sigma(\overleftarrow{W}_o \cdot m_t + \overleftarrow{U}_o \cdot \overleftarrow{h}_{t+1} + \overleftarrow{b}_o)$$
$$\overleftarrow{z} = \tanh(\overleftarrow{W}_z \cdot m_t + \overleftarrow{U}_z \cdot \overleftarrow{h}_{t+1} + \overleftarrow{b}_z)$$

As is often done with bidirectional LSTM chains, all weights of the parameter matrices in the forward chain $\overrightarrow{W}, \overrightarrow{U}, \overrightarrow{b}$ are shared among all cells of the forward chain, and all weights of the parameter matrices in the backward chain $\overleftarrow{W}, \overleftarrow{U}, \overleftarrow{b}$ are shared among all cells of the backward chain. As in Mollá (2017a)'s framework, there are separate sets of parameter matrices for the sentence and for the question.

Mollá (2017a) used all the sentences of the full abstracts as the candidate input. Given that subsequent experiments observed an improvement of results by using the snippets only, our entries to BioASQ 6b used the snippets only. Also, Mollá (2017a) observed very competitive results of a simple baseline that returned the first n sentences. This suggests that sentence position is a useful feature and we consequently incorporated the sentence position as a feature in stage 3.

Given the sentence embedding s_i and question embedding q_i, each obtained either by the mean of embeddings or by applying LSTM chains as described above, and given the sentence position p_i, stage 3 is implemented as a simple neural network with one hidden layer with a relu activation, followed by a linear combination:

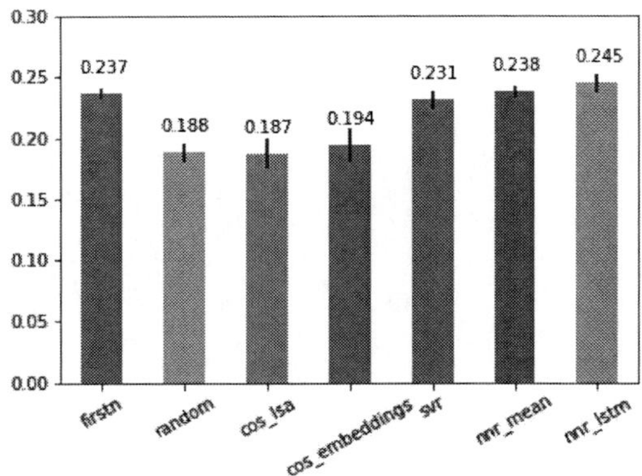

Figure 1: Comparison of deep learning experiments with several baselines. The error bars indicate the standard deviation of 10-fold cross-validation.

$$r_i = \max(0, W_r\,[s_i; q_i \odot s_i; p_i] + b_r)$$
$$score_i = W_{score} r_i + b_{score}$$

Following Mollá (2017a)'s framework, we used the element-wise multiplication between q_i and s_i as a way to encode the similarity between the question and the input sentence.

Figure 1 compares the results of the deep learning approaches against the following baselines:

Firstn. Return the first n sentences. As mentioned above, this baseline is often rather hard to beat.

Random. Return n random sentences. This is the lower bound of any extractive summarisation system.

Cos LSA. Return the n sentences that have the highest cosine similarity with the question. The feature vectors for the computation of the cosine similarity were obtained by computing $tf.idf$, followed by a dimension reduction step that selected the top 200 components after Latent Semantic Analysis.

Cos Embeddings. Return the n sentences that have the highest cosine similarity with the question. The cosine similarity was based on the sum of the word embeddings (using embeddings with 200 dimensions) in the sentence/question.

SVR. Train a Support Vector Regression system that uses as features a combination of $tf.idf$,

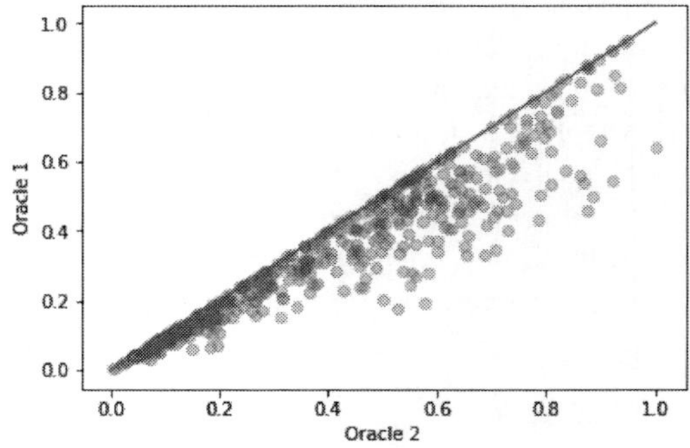

Figure 2: Scatter-plot comparing the F1 ROUGE-SU4 score of two oracles using a random sample of 500 questions.

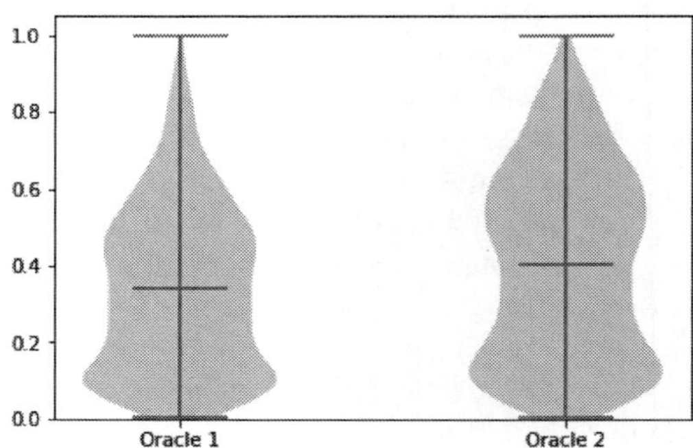

Figure 3: Violin plots of the F1 ROUGE-SU4 scores of two oracles.

word embeddings and distance metrics as described by Mollá (2017a), plus the position of the snippet.

Figure 1 shows that the "firstn" baseline is indeed hard to beat, and was matched only by the deep learning frameworks. Of these, the system using LSTM obtained the best results and was chosen for submission to BioASQ.

3 Reinforcement Learning

While the results using deep learning are encouraging, the models are trained on individually annotated sentences, and a summary is obtained by selecting the top k sentences. An upper bound of the results obtained using such an approach would be an oracle that selects the k sentences with highest individual ROUGE scores. Figures 2 and 3 show a comparison between the following two oracles:

Oracle 1. Return the k snippets with highest *in-*

dividual ROUGE score. This is a reasonable upper bound of supervised machine learning approaches such as presented in Section 2.

Oracle 2. Return the combination of k snippets with highest *collective* ROUGE score. In particular, the oracle calculates the ROUGE score of every combination of k snippets, and selects the combination with highest ROUGE score. This is an upper bound of any conceivable extractive summarisation system that returns k snippets.

Figure 2 shows the scatter-plot between oracles 1 and 2. It shows that oracle 1 under-performs oracle 2 in a number of questions. Figure 3 plots the distributions of the ROUGE scores of each oracle side by side, and it clearly shows that the mean of the ROUGE scores of oracle 1 is lower than that of oracle 2.

Reinforcement learning allows to train the system on the ROUGE score of the final summary. This is done by iteratively allowing the system to extract a summary based on its current policy, and then updating the policy based on the feedback given by the ROUGE score of the extracted summary.

The reinforcement learning approach in our system is based on Mollá (2017b)'s method. In particular, the reinforcement learning agent receives as input a candidate sentence and additional context information, and uses a global policy to determine the best possible action (either to select the sentence or not to select it). The global policy is implemented as a neural network with a hidden layer and is trained on a training partition of the development data by applying a variant of REINFORCE (Williams, 1992).

More specifically, the global policy learns a set of parameters θ so that the policy predicts the probability of not selecting the sentence ($Pr(a = 0; \theta)$) by applying the following neural network:

$$\begin{aligned} Pr(a = 0; \theta) &= \sigma(W_h h + b_h) \\ h &= \max(0, W_s s + b_s) \end{aligned}$$

where $\theta = [W_h; W_s; b_h; b_s]$ and the input h is the concatenation of the following features:

1. *tf.idf* of candidate sentence i;

2. *tf.idf* of the entire input text to summarise;

3. *tf.idf* of the summary generated so far;

4. *tf.idf* of the candidate sentences that are yet to be processed;

5. *tf.idf* of the question; and

6. Length (in number of sentences) of the summary generated so far.

The features chosen are such that the global policy has information about the candidate sentence (1), the entire list of candidate sentences (2), the summary that has been generated so far (3), the input sentences that are yet to be processed (4), and the question (5). Experiments by Mollá (2017b) show that this information suffices to learn a global policy. In addition, we added the length of the summary generated so far (6). Our preliminary experiments showed that this additional feature facilitated a faster learning of the policy, and produced better results overall.

The specific algorithm that learns the global policy is presented in Figure 1. The system

Data: `train_data`
Result: θ

```
 1  sample ~ Uniform(train_data);
 2  s ← env.reset(sample);
 3  all_gradients ← ∅;
 4  initialise(θ);
 5  episode ← 0;
 6  while True do
 7  |   ξ ~ Bernoulli ( (Pr(a=0;θ)+p) / (1+2×p) );
 8  |   y ← 1 − ξ;
 9  |   gradient ←
    |       ∇(cross_entropy(y,Pr(a=0;θ))) / ∇θ ;
10  |   all_gradients.append(gradient);
11  |   s, r, done ← env.step(ξ);
12  |   episode ← episode + 1;
13  |   if done then
14  |   |   θ ←
    |   |       θ − α × r × mean(all_gradients);
15  |   |   sample ~ Uniform(train_data);
16  |   |   s ← env.reset(sample);
17  |   |   all_gradients ← ∅;
18  |   end
19  end
```

Algorithm 1: Training by Policy Gradient, where $\theta = [W_h; W_s; b_h; b_s]$.

first chooses one question from the training data (line 1). Then, after randomly initialising the parameters of the global policy (line 4), it iteratively

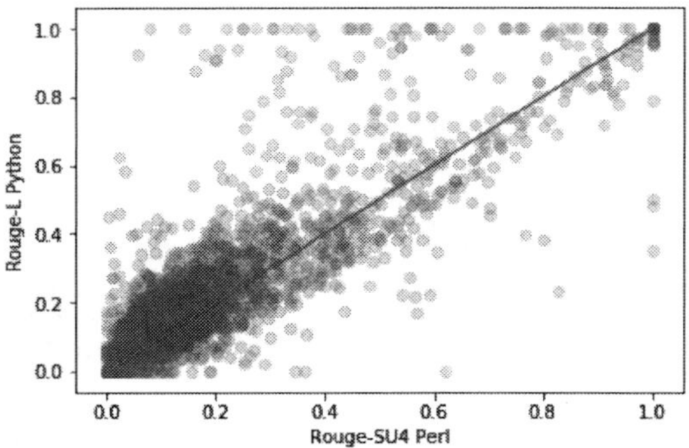 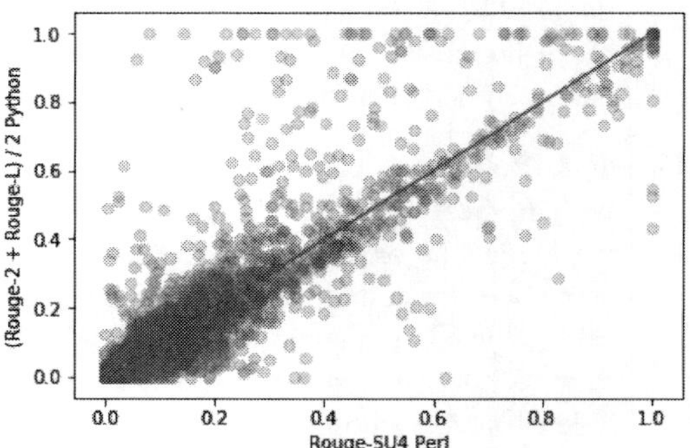

Figure 4: Comparison between the Python and Perl implementations of ROUGE for a random sample of 5000 snippets and their respective ideal answers. The Python implementation uses ROUGE-L. The Perl implementation uses ROUGE-SU4.

Figure 5: Comparison between the Python and Perl implementations of ROUGE for a random sample of 5000 snippets and their respective ideal answers. The Python implementation uses (ROUGE-2 + ROUGE-L) / 2. The Perl implementation uses ROUGE-SU4.

samples an action from the current global policy plus some perturbation p (line 7) and applies the action (line 11). When all the candidate sentences related to the question have been processed and actioned on (line 13), the resulting summary is evaluated and a reward r produced (done previously, in line 11). Then, the policy parameters are updated by multiplying the mean of all gradients obtained at every step by the reward (line 14), and a new question is selected from the training data (line 15). Each iteration step is called an episode (lines 5 and 12).

The perturbation p forces a wide exploration of the action space during the first episodes and is gradually reduced at every episode according to this formula:

$$p = 0.2 \times 3000/(3000 + \text{episode})$$

3.1 ROUGE Variants

The evaluation scripts of the BioASQ task used the original Perl implementation of ROUGE-2 and ROUGE-SU4 (Lin, 2004). Our experiments aimed to use ROUGE-SU4. Whereas the Perl implementation of ROUGE was used for the deep learning experiments described in section 2, as an implementation decision we used Python's `pyrouge` library for the reinforcement learning experiments. Python's `pyrouge` provides ROUGE-1, ROUGE-2 and ROUGE-L, but not ROUGE-SU4.

Figures 4 and 5 compare the ROUGE F1 scores of the Python libraries against the ROUGE-SU4 F1 score of the Perl implementation. Figure 4

Kernel	C	gamma
rbf	1.0	0.1

Table 2: Settings used in the SVR experiments.

uses the Python library for ROUGE-L, and Figure 5 uses the Python library for the mean between ROUGE-2 and ROUGE-L. We can observe some noise in the correlation between the Perl and Python implementations, but in general the mean between ROUGE-2 and ROUGE-L was a better approximation of the Perl implementation of ROUGE-SU4.

In general, we observed lower results of the Python versions of ROUGE-L and ROUGE-2 compared with the Perl versions. On the light of this, we strongly advise always to specify the implementation of ROUGE being used, since the results produced by different versions may not be comparable.

4 Settings

The snippets were split into sentences using NLTK's sentence tokeniser.

The $tf.idf$ information used for the baselines was computed using NLTK's TfidfVectorizer. As in the system by Mollá (2017a), this vectoriser was trained using a collection of text consisting of the questions and the ideal answers.

The SVR experiments were implemented using Python's `sklearn` library and used word embeddings with 200 dimensions. The specific settings of the SVR model are shown in Table 2.

The deep learning experiments were imple-

System	Batch	Dropout	Epochs
Mean	4096	0.4	5
LSTM	4096	0.8	10

Table 3: Settings used in the deep learning experiments.

mented using TensorFlow's libraries. The specific details of the model are:

- Dimension of embeddings: 100

- Length of the LSTM chain: 300. Sentences with more than 300 words were truncated.

- Dimension of $\overrightarrow{h}$ and $\overleftarrow{h}$: 100

- Number of cells in the final hidden layer: 50

Table 3 shows the training settings used by the deep learning architectures. These settings were obtained empirically in a fine-tuning stage.

The reinforcement learning experiments were implemented in TensorFlow. Due to hardware constraints, the reinforcement learning approach only processed the first 20 sentences. The specific details of the architecture of the global policy are:

- Number of cells in the hidden layer: 200

The global policy was trained using a training partition and evaluated on a separate partition. The global policy parameters that generated the best results in the evaluation partition were used for the final runs to BioASQ.

5   Results

We submitted 5 runs for each batch as listed below.

MQ-1: Return the first n sentences. This is the Firstn baseline described in Section 2.

MQ-2: Return the n sentences with highest cosine similarity with the question. This is the Cos Embeddings baseline described in Section 2.

MQ-3: Return the n sentences according to the SVR baseline described in Section 2.

MQ-4: Score the sentences using the LSTM-based deep learning approach as described in Section 2.

MQ-5: Apply reinforcement learning as described in Section 3, with the variations described below.

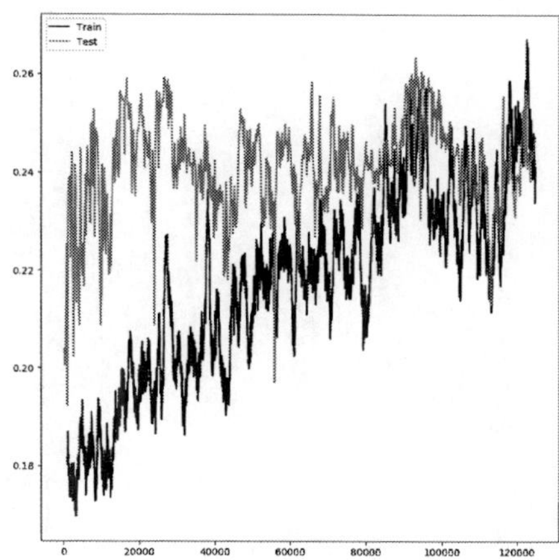

Figure 6: Reinforcement Learning model trained for batch 1. The reward (y axis) is ROUGE-L.

The policy trained for the reinforcement learning approach of run MQ-5 varied in several of the batches. In particular, the first batch was trained on ROUGE-L while batches 2 to 5 were trained on (ROUGE-2 + ROUGE-L) / 2. Also, to test the impact of different initialisation settings, we trained the system twice and generated two separate models. Batch 2 used one model, whereas batches 3 to 5 used the other model.

Figures 6 and 7 show the evolution of the results of the policies during the training stage for batches 1 and 2. We observe some differences during training, but in general the best model on the test set achieved a ROUGE score between 0.25 and 0.26.[2] This is higher than the results reported by Mollá (2017b), who reported a ROUGE score of about 0.2. The likely cause of the improvement in the results is the inclusion of the length of the summary generated so far in the context available by the policy.

Figure 8 shows the results of the submissions to BioASQ. In general, the deep learning approach (MQ-4) achieved the best results. While the "first n" baseline (MQ-1) was fairly competitive and outperformed some of the runs of other participants to BioASQ, the baseline was not as strong as reported by Mollá (2017a) on BioASQ5b.

We can also observe that the evaluation re-

---

[2]The training stage for batches 3 to 5 achieved a best result slightly over 0.26.

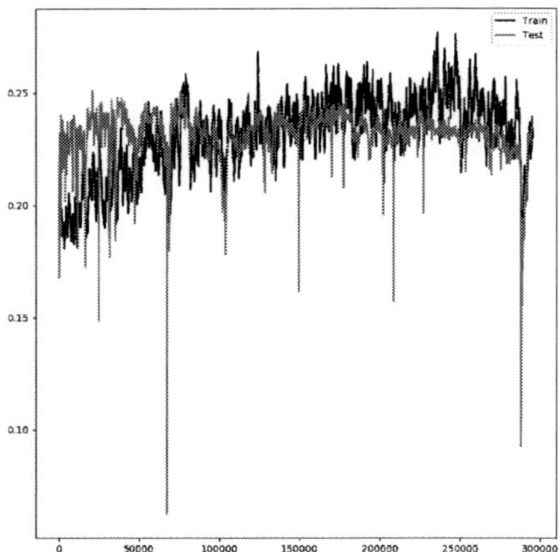

Figure 7: Reinforcement Learning model trained for batch 2. The reward (y axis) is (ROUGE-2+ROUGE-L)/2.

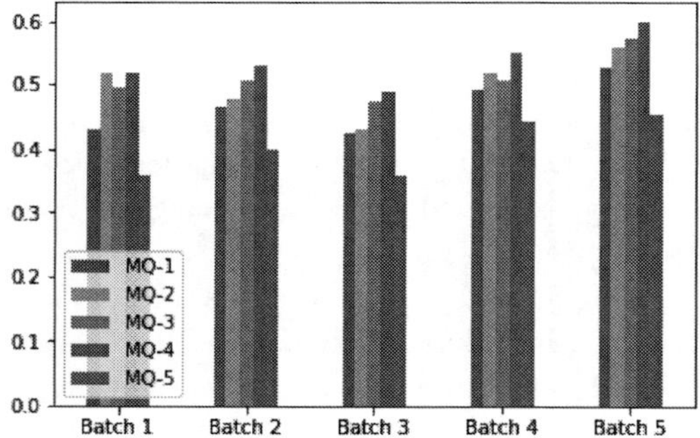

Figure 8: ROUGE-SU4 results of the BioASQ runs.

sults of the submissions to BioASQ give higher ROUGE scores than those obtained in our experiments, both using the original Perl implementation of ROUGE, and the Python version. In fact, all of the runs except for MQ-5 reported better BioASQ results than our experiments with the oracles. This is worth investigating.

The runs using reinforcement learning (MQ-5) gave worse results than the other runs. The cause for this is also worth investigating, especially considering that, in our experiments, the results of the reinforcement learning approach were very competitive compared with the results of the other approaches.

6 Conclusions

In this paper we have described the deep learning and reinforcement learning approaches used for the runs submitted to BioASQ 6b, phase B, for the generation of ideal answers.

The deep learning approach used a supervised regression set-up to score the individual candidate sentences. The training data was generated by computing the ROUGE score of each individual candidate sentence, and the summary was obtained by selecting the top-scoring sentences. The results of the deep learning runs outperformed all other runs.

The reinforcement learning approach trained a global policy using as a reward the ROUGE score of the summary generated by the policy. The results of our experiments were very competitive but the submission results were lower than those of the other runs.

Further work will focus on the refinement of the reinforcement learning approach. In particular, further work will include the addition of a baseline in the training of the policy, as it has been shown to reduce the variance and to speed up the training process. Also, the architecture of the neural network implementation of the policy will be revised and enhanced by incorporating a more sophisticated model.

Further work on the deep learning runs will focus on the incorporation of more complex models. For example, preliminary experiments seem to indicate that a classification-based approach could outperform the current regression-based approach. Also, it is expected that a sequence-labelling approach would produce better results since the candidate sentences would not be processed independently.

References

Sepp Hochreiter, Jrgen Schmidhuber, Sepp Hochreiter, Jrgen Schmidhuber, and Jrgen Schmidhuber. 1997. Long short-term memory. *Neural Computation*, 9(8):1735–80.

Chin-Yew Lin. 2004. ROUGE: A package for automatic evaluation of summaries. In *ACL Workshop on Tech Summarisation Branches Out*.

Tomas Mikolov, Kai Chen, Greg Corrado, and Jeffrey Dean. 2013. Efficient estimation of word representations in vector space. In *Proceedings of Workshop at ICLR*, pages 1–12.

Diego Mollá. 2017a. Macquarie University at BioASQ 5b — query-based summarisation techniques for selecting the ideal answers. In *Proc. BioNLP2017*.

Diego Mollá. 2017b. Towards the use of deep reinforcement learning with global policy for query-based extractive summarisation. In *Proceedings of the Australasian Language Technology Association Workshop 2017*, pages 103–107.

George Tsatsaronis, Georgios Balikas, Prodromos Malakasiotis, Ioannis Partalas, Matthias Zschunke, Michael R Alvers, Dirk Weissenborn, Anastasia Krithara, Sergios Petridis, Dimitris Polychronopoulos, Yannis Almirantis, John Pavlopoulos, Nicolas Baskiotis, Patrick Gallinari, Thierry Artiéres, Axel-Cyrille Ngonga Ngomo, Norman Heino, Eric Gaussier, Liliana Barrio-Alvers, Michael Schroeder, Ion Androutsopoulos, and Georgios Paliouras. 2015. An overview of the BIOASQ large-scale biomedical semantic indexing and question answering competition. *BMC Bioinformatics*, 16(1):138.

Ronald J. Williams. 1992. Simple statistical gradient-following algorithms for connectionist reinforcement learning. *Machine Learning*, 8:229–256.

AUEB at BioASQ 6: Document and Snippet Retrieval

**Georgios-Ioannis Brokos[1], Polyvios Liosis[1], Ryan McDonald[1,3],
Dimitris Pappas[1,2] and Ion Androutsopoulos[1]**

[1]Dept. of Informatics, Athens University of Economics and Business, Greece
[2]Institute for Language and Speech Processing, Research Center 'Athena', Greece
[3]Google AI

Abstract

We present AUEB's submissions to the
BioASQ 6 document and snippet retrieval
tasks (parts of Task 6b, Phase A). Our models use novel extensions to deep learning architectures that operate solely over the text of the
query and candidate document/snippets. Our
systems scored at the top or near the top for
all batches of the challenge, highlighting the
effectiveness of deep learning for these tasks.

1 Introduction

BioASQ (Tsatsaronis et al., 2015) is a biomedical
document classification, document retrieval, and
question answering competition, currently in its
sixth year.[1] We provide an overview of AUEB's
submissions to the document and snippet retrieval
tasks (parts of Task 6b, Phase A) of BioASQ 6.[2]
In these tasks, systems are provided with English
biomedical questions and are required to retrieve
relevant documents and document snippets from a
collection of MEDLINE/PubMed articles.[3]

We used deep learning models for both
document and snippet retrieval. For document retrieval, we focus on extensions to the
Position-Aware Convolutional Recurrent Relevance (PACRR) model of Hui et al. (2017) and,
mostly, the Deep Relevance Matching Model
(DRMM) of Guo et al. (2016), whereas for snippet retrieval we based our work on the Basic Bi-
CNN (BCNN) model of Yin et al. (2016). Little
task-specific pre-processing is employed and the
models operate solely over the text of the query
and candidate document/snippets.

Overall, our systems scored at the top or near
the top for all batches of the challenge. In previous

---

[1]Consult http://bioasq.org/.

[2]For further information on the BioASQ 6 tasks, see
http://bioasq.org/participate/challenges.

[3]http://www.ncbi.nlm.nih.gov/pubmed/.

years of the BioASQ challenge, the top scoring
systems used primarily traditional IR techniques
(Jin et al., 2017). Thus, our work highlights that
end-to-end deep learning models are an effective
approach for retrieval in the biomedical domain.

2 Document Retrieval

For document retrieval, we investigate new deep
learning architectures focusing on *term-based interaction models*, where query terms (*q-terms* for
brevity) are scored relative to a document's terms
(*d-terms*) and their scores are aggregated to produce a relevance score for the document. All models use pre-trained embeddings for all q-terms and
d-terms. Details on data resources and data pre-
processing are given in Section 5.1.

2.1 PACRR-based Models

The first model we investigate is PACRR (Hui et al.,
2017). In this model, a query-document term similarity matrix *sim* is first computed (Fig. 1, left).
Each cell (i, j) of *sim* contains the cosine similarity between the embeddings of a q-term q_i and
a d-term d_j. To keep the dimensions $l_q \times l_d$ of
sim fixed across queries and documents of vary-
ing lengths, queries are padded to the maximum
number of q-terms l_q, and only the first l_d terms
per document are retained.[4] Then, convolutions of
different kernel sizes $n \times n$ ($n = 2, \ldots, l_g$) are
applied to *sim* to capture n-gram query-document
similarities. For each size $n \times n$, multiple kernels (filters) are used. Max pooling is then applied
along the dimension of the filters (max value of all
filters of the same size), followed by k-max pooling along the dimension of d-terms to capture the
strongest k signals between each q-term and all
the d-terms. The resulting matrices (one per kernel

---

[4]We use PACRR-*firstk*, which Hui et al. (2017) recommend
when documents fit in memory, as in our experiments.

30

Proceedings of the 2018 EMNLP Workshop BioASQ: Large-scale Biomedical Semantic Indexing and Question Answering, pages 30–39
Brussels, Belgium, November 1st, 2018. ©2018 Association for Computational Linguistics

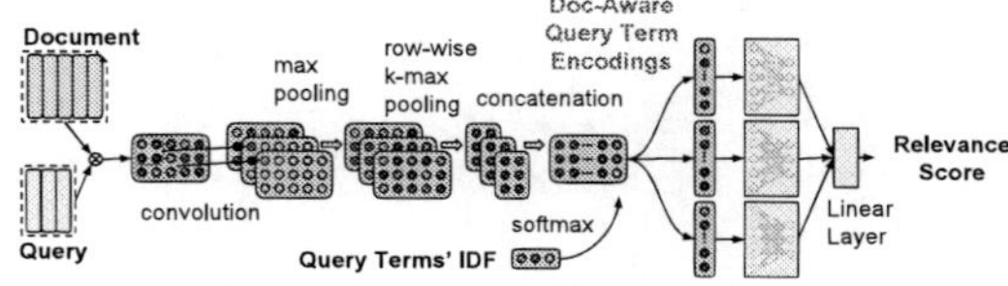

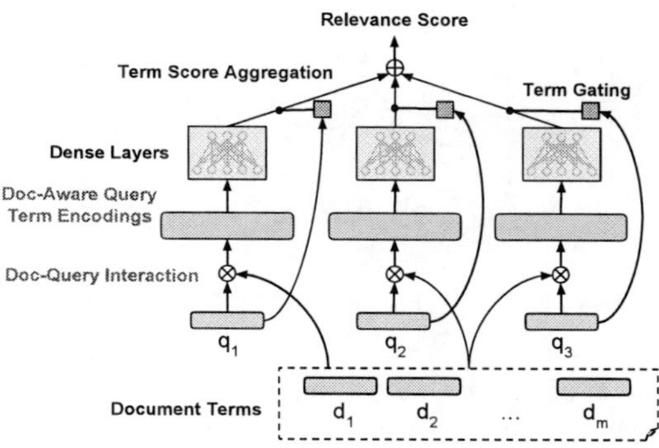

Figure 1: PACRR (Hui et al., 2017) and TERM-PACRR. In PACRR, an MLP is applied to the concatenation of the document-aware q-term encodings to produce the relevance score. In TERM-PACRR, the MLP is applied *separately* to each document-aware q-term encoding; the resulting scores are combined by a linear layer.

size) are concatenated into a single matrix where each row is a document-aware q-term encoding (Fig. 1); the IDF of the q-term is also appended, normalized by applying a softmax across the IDFs of all the q-terms. Following Hui et al. (2018), we concatenate the rows of the resulting matrix into a single vector, which is passed to an MLP that produces a query-document relevance score.[5]

Instead of using an MLP to score the *concatenation* of all the (document-aware) q-term encodings, a simple extension we found effective was to use an MLP to *independently* score each q-term encoding (the same MLP for all q-terms, Fig. 1); the resulting scores are aggregated via a linear layer. This version, TERM-PACRR, performs better than PACRR, using the same number of hidden layers in the MLPs. Likely this is due to the fewer parameters of TERM-PACRR's MLP, which is shared across the q-term representations and operates on shorter input vectors. Indeed, in our early experiments TERM-PACRR was less prone to overfitting.[6]

2.2 DRMM-based Models

The second model we investigate is DRMM (Guo et al., 2016) (Fig. 2). The original DRMM uses pretrained word embeddings for q-terms and d-terms, and (bucketed) cosine similarity histograms (outputs of $\otimes$ nodes in Fig. 2). Each histogram captures the similarity of a q-term to all the d-terms of a particular document. The histograms, which in this model are the document-aware q-term encodings, are fed to an MLP (dense layers of Fig. 2) that produces the (document-aware) score of each q-term. Each q-term score is then weighted using

Figure 2: Illustration of DRMM (Guo et al., 2016) for three q-terms and m d-terms. The $\otimes$ nodes produce (bucketed) cosine similarity histograms, each capturing the similarity between a q-term and all the d-terms.

a gating mechanism (topmost box nodes in Fig. 2) that examines properties of the q-term to assess its importance for ranking (e.g., common words are less important). The sum of the weighted q-term scores is the relevance score of the document.

For gating (topmost box nodes of Fig. 2), Guo et al. (2016) use a linear self-attention:

$$g_i = \mathrm{softmax}\left(w_g^T\,\phi_g(q_i); q_1, \ldots, q_n\right)$$

$\phi_g(q_i)$ is the embedding $e(q_i)$ of the i-th q-term, or its IDF, $\mathrm{idf}(q_i)$; w_g is a weights vector. We found that $\phi_g(q_i) = [e(q_i); \mathrm{idf}(q_i)]$, where ';' is concatenation, was optimal for all DRMM-based models.

2.2.1 ABEL-DRMM

The original DRMM (Guo et al., 2016) has two shortcomings. The first one is that it ignores entirely the contexts where the terms occur, in contrast to position-aware models such as PACRR (Section 2.1) or those based on recurrent representations (Palangi et al., 2016). Secondly, the histogram representation for document-aware q-term encodings is not differentiable, so it is not possible to train the network end-to-end, if one wished to backpropagate all the way to word embeddings.

To address the first shortcoming, we add an encoder (Fig. 3) to produce the *context-sensitive encoding* of each q-term or d-term from the pretrained embeddings of the previous, current, and next term in a particular query or document. A single dense layer with residuals is used, in effect a one-layer Temporal Convolutional Network (TCN) (Bai et al., 2018) without pooling or dilation. The number of convolutional filters equals the dimen-

---

[5]Hui et al. (2017) used an additional LSTM, which was later replaced by the final concatenation (Hui et al., 2018).

[6]In the related publication of McDonald et al. (2018) TERM-PACRR is identical to the PACRR-DRMM model.

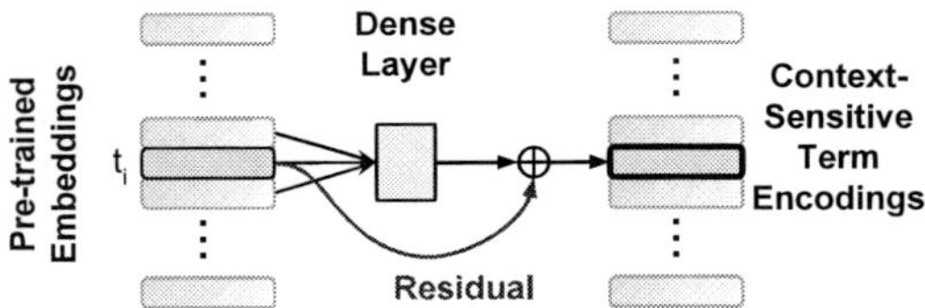

Figure 3: Producing *context-sensitive term encodings*.

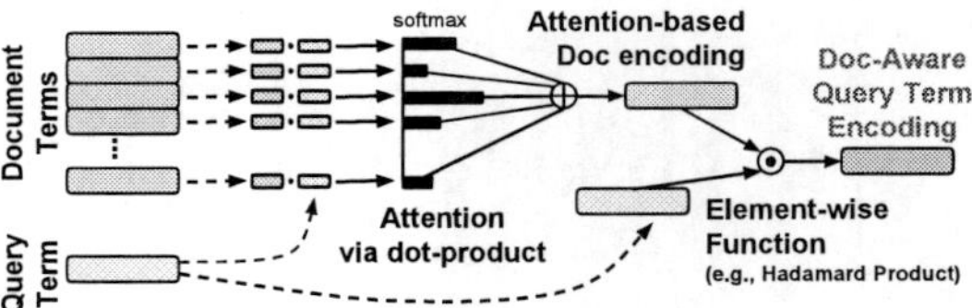

Figure 4: ABEL-DRMM sub-net. From context-aware q-term and d-term encodings (Fig. 3), it generates fixed-dimension *document-aware q-term encodings* to be used in DRMM (Fig. 2, replacing $\otimes$ nodes).

sions of the pre-trained embedding, for residuals to be summed without transformation.

Specifically, let $e(t_i)$ be the pre-trained embedding for a q-term or d-term term t_i. We compute the context-sensitive encoding of t_i as:

$$c(t_i) = \varphi\Big(W_c\, \phi_c(t_i) + b_c\Big) + e(t_i) \qquad (1)$$

W_c and b_c are the weights matrix and bias vector of the dense layer, φ is the activation function, $\phi_c(t_i) = [e(t_{i-1}); e(t_i); e(t_{i+1})]$, t_{i-1}, t_{i+1} are the tokens surrounding t_i in the query or document. This is an orthogonal way to incorporate context into the model relative to PACRR. PACRR creates a query-document similarity matrix and computes n-gram convolutions over the matrix. Here we incorporate context directly into the term encodings; hence similarities in this space are already context-sensitive. One way to view this difference is the point at which context enters the model – directly during term encoding (Fig. 3) or after term similarity scores have been computed (PACRR, Fig. 1).

To make DRMM trainable end-to-end, we replace its histogram-based document-aware q-term encodings ($\otimes$ nodes of Fig. 2) by q-term encodings that consider d-terms via an attention-mechanism. Figure 4 shows the new sub-network that computes the document-aware encoding of a q-term q_i, given a document $d = \langle d_1, \dots, d_m \rangle$ of m d-terms. We first compute a dot-product attention score $a_{i,j}$ for each d_j relative to q_i:

$$a_{i,j} = \text{softmax}\Big(c(q_i)^T c(d_j); d_1, \dots, d_m\Big) \qquad (2)$$

where $c(t)$ is the context-sensitive encoding of t (Eq. 1). We then sum the context-sensitive encodings of the d-terms, weighted by their attention scores, to produce an attention-based representation d_{q_i} of document d from the viewpoint of q_i:

$$d_{q_i} = \sum_j a_{i,j}\, c(d_j) \qquad (3)$$

The Hadamard product (element-wise multiplication, $\odot$) between the document representation d_{q_i}

and the q-term encoding $c(q_i)$ is then computed and used as the fixed-dimension document-aware encoding $\phi_H(q_i)$ of q_i (Fig. 4):

$$\phi_H(q_i) = d_{q_i} \odot c(q_i) \qquad (4)$$

The $\otimes$ nodes and lower parts of the DRMM network of Fig. 2 are now replaced by (multiple copies of) the sub-network of Fig. 4 (one copy per q-term), with the $\odot$ nodes replacing the $\otimes$ nodes. We call the resulting model Attention-Based Element-wise DRMM (ABEL-DRMM).

Intuitively, if the document contains one or more terms d_j that are similar to q_i, the attention mechanism will have emphasized mostly those terms and, hence, d_{q_i} will be similar to $c(q_i)$, otherwise not. This similarity could have been measured by the cosine similarity between d_{q_i} and $c(q_i)$, but the cosine similarity assigns the same weight to all the dimensions, i.e., to all the element-wise products in $\phi_H(q_i)$. By using the Hadamard product, we pass on to the upper layers of DRMM (the dense layers of Fig. 2), which score each q-term with respect to the document, all the element-wise products of $\phi_H(q_i)$, allowing the upper layers to learn which element-wise products (or combinations of them) are important when matching a q-term to the document.

2.2.2 ABEL-DRMM extensions

We experimented with two extensions to ABEL-DRMM. The first is a *density-based extension* that considers all the windows of l_w consecutive tokens of the document and computes the ABEL-DRMM relevance score per window. The final relevance score of a document is the sum of the original ABEL-DRMM score computed over the entire document plus the maximum ABEL-DRMM score over all the document's windows. The intuition is to reward not only documents that match the query, but also those that match it in a dense window.

The second extension is to compute a *confidence* score per document and only return documents with scores above a threshold. We apply a softmax over the ABEL-DRMM scores of the top t_d documents and return only documents from the top t_d with normalized scores exceeding a threshold t_c. While this will always hurt metrics like Mean Average Precision (MAP) when evaluating document retrieval, it has the potential to improve the precision of downstream components, in our case snippet retrieval, which in fact we observe.

3   Snippet Retrieval

For the snippet retrieval task, we used the 'basic CNN' (BCNN) network of the broader ABCNN model (Yin et al., 2016), which we combined with a post-processing stage, as discussed below. The input of snippet retrieval is an English question and text snippets (e.g., sentences) from documents that the document retrieval component returned as relevant to the question. The goal is to rank the snippets, so that snippets that human experts selected as relevant to the question will be ranked higher than others. In BioASQ, human experts are instructed to select relevant snippets consisting of one or more consecutive sentences.[7] For training purposes, we split the relevant documents into sentences, and consider sentences that overlap the gold snippets (the ones selected by the human experts) as relevant snippets, and the remaining ones as irrelevant. At inference time, documents returned by the document retrieval model as relevant are split into sentences, and these sentences are ranked by the system. For details on sentence splitting, tokenization, etc., see Section 5.1.

3.1   BCNN Model

BCNN receives as input two sequences of terms (tokens), in our case a question (query) and a sentence from a document. All terms are represented by pre-trained embeddings (Section 5.1). Snippet sequences were truncated (or zero padded) to be of uniform length. A convolution layer with multiple filters, each of the same width w, is applied to each one of the two input sequences, followed by a windowed-average pooling layer over the same filter width to produce a feature map (per filter) of the same dimensionality as the input to the con-

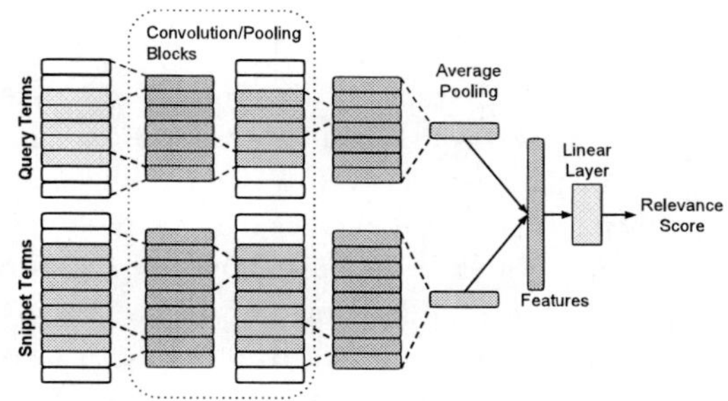

Figure 5: BCNN (Yin et al., 2016) scoring snippets relative to a query. The example illustrates a query of 5 terms, a snippet of 7 terms, and a single convolution filter of width $w = 3$. Zero-padding shown as empty boxes. In each convolution/pooling block, the convolution layer is followed by a windowed-average pooling of the same width w to preserve the dimensionality of the input to the block. Thus convolution/pooling blocks can be repeated, making the model arbitrarily deep.

volution layer.[8] Consequently, we can stack an arbitrary number of convolution/pooling blocks in order to extract increasingly abstract features.

An average pooling layer is then applied to the entire output of the last convolution/pooling block (Fig. 5) to obtain a feature vector of the query and snippet, respectively. When multiple convolution filters are used (Fig. 5 illustrates only one), we obtain a different feature vector from each filter (for the query and snippet, respectively), and the feature vectors from the different filters are concatenated, again obtaining a single feature vector for the query and snippet, respectively. Similarity scores are then computed from the query and snippet feature vectors, and these are fed into a linear logistic regression layer. One critical implementation detail from the original BCNN paper is that when computing the query-snippet similarity scores, average pooling is actually applied to the output of each one of the convolution/pooling blocks, i.e., we obtain a different query and snippet feature vector from the output of each block. Different similarity scores are computed based on the query and snippet feature vectors obtained from the output of each block, and all the similarity scores are passed to the final layer. Thus the number of inputs to the final layer is proportional to the number of blocks.

---

[7]This was not actually the case in BioASQ year 1. Hence, some of our training data do not adhere to this rule.

[8]The same filters are applied to both queries and snippets.

3.2 Post-processing

A technique that seems to improve our results in snippet retrieval is to retain only the top K_s snippets with the best BCNN scores for each query, and then re-rank the K_s snippets by the relevance scores of the documents they came from; if two snippets came from the same document, they are subsequently ranked by their BCNN score. This is a proxy for more sophisticated models that would jointly consider document and snippet retrieval. This is important as the snippet retrieval model is trained under the condition that it only sees relevant documents. So accounting for the rank/score of the document itself helps to correctly bias the snippet model.

4 Overall System Architecture

Figure 6 outlines the general architecture that we used to piece together the various components. It consists of retrieving the top N documents per query using BM25 (Robertson et al., 1995); re-ranking the top N documents using one of the document retrieval models (Section 2) and retaining (up to) the top K_d documents; scoring all candidate snippets of the top K_d documents via a snippet retrieval model (BCNN, Section 3.1) and retaining (up to) the top K_s snippets; re-ranking the K_s snippets by the relevance scores of the documents they came from (Section 3.2).[9]

We set $K_d = K_s = 10$ as it was dictated by the BioASQ challenge. We set $N = 100$ as we found that with this value, BM25 returned the majority of the relevant documents from the training/development data sets. Setting N to larger values had no impact on the final results. The reason for using a pre-retrieval model based on BM25 is that the deep document retrieval models we use here are computationally expensive. Thus, running them on every document in the index for every query is prohibitive, whereas running them on the top $N = 100$ documents from a pre-retrieval system is easily achieved.

5 Experiments

All retrieval components (PACRR-, DRMM-, BCNN-based) were augmented to combine the scores of the corresponding deep model with a number of traditional IR features, which is a common technique (Severyn and Moschitti, 2015). In

---

[9]The last step was used only in batches 3–5.

TERM-PACRR, the additional features are fed to the linear layer that combines the q-term scores (Fig. 1). In ABEL-DRMM, an additional linear layer is used that concatenates the deep learning document relevance score with the traditional IR features. In BCNN, the additional features are included in the final linear layer (Fig. 5). The additional features we used were the BM25 score of the document (the document the snippet came from, in snippet retrieval), word overlap (binary and IDF weighted) between the query and the document or snippet; bigram overlap between the query and the document or snippet. The latter features were taken from Mohan et al. (2017). The additional features improved the performance of all models.

5.1 Data Resources and Pre-processing

The document collection consists of approx. 28M 'articles' (titles and abstracts only) from the 'MEDLINE/PubMed Baseline 2018' collection.[10] We discarded the approx. 10M articles that contained only titles, since very few of these were annotated as relevant. For the remaining 18M articles, a document was the concatenation of each title and abstract. These documents were then indexed using Galago, removing stop words and applying Krovetz's stemmer (Krovetz, 1993).[11] This served as our pre-retrieval model.

Word embeddings were pre-trained by applying word2vec (Mikolov et al., 2013) to the 28M 'articles' of the MEDLINE/PubMed collection. IDF values were computed over the 18M articles that contained both titles and abstracts. We used the GenSim implementation of word2vec (skip-gram model), with negative sampling, window size set to 5, default other hyper-parameter values, to produce word embeddings of 200 dimensions.[12] The word embeddings were not updated when training the document relevance ranking models. For tokenization, we used the 'bioclean' tool provided by BioASQ.[13] In snippet retrieval, we used NLTK's

---

[10]Available from `https://www.nlm.nih.gov/databases/download/pubmed_medline.html`.

[11]We used Galago version 3.10. Consult `http://www.lemurproject.org/galago.php`.

[12]Consult `https://radimrehurek.com/gensim/models/word2vec.html`. We used Gensim v. 3.3.0. The word embeddings and code of our experiments are available at `https://github.com/nlpaueb/aueb-bioasq6`.

[13]The tool accompanies an older set of embeddings provided by BioASQ. See `http://participants-area.bioasq.org/tools/BioASQword2vec/`.

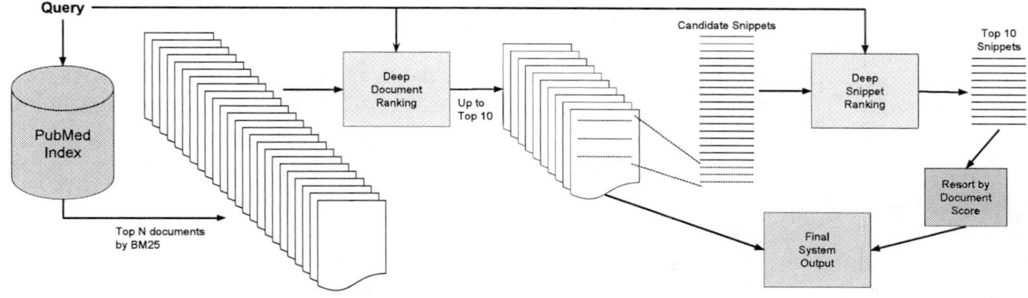

Figure 6: Overall architecture of document and snippet retrieval systems.

English sentence splitter.[14]

To train and tune the models we used years 1–5 of the BioASQ data, using batch 5 of year 5 as development for the final submitted models, specifically when selecting optimal model epoch. We report test results (F1, MAP, GMAP) on batches 1–5 of year 6 from the official results table.[15] Details on the three evaluation metrics are provided by Tsatsaronis et al. (2015). They are standard, with the exception that MAP here always assumes 10 relevant documents/snippets, which is the maximum number of documents/snippets the participating systems were allowed to return per query.

5.2 Hyperparameters

All DRMM-based models were trained with Adam (Kingma and Ba, 2014) with a learning rate of 0.01 and $\beta_1/\beta_2 = 0.9/0.999$. Batch sizes were set to 32. We used a hinge-loss with a margin of 1.0 over pairs of a single positive and a single negative document of the same query. All models used a two-layer MLP to score q-terms (dense layers of Fig. 2), with leaky-RELU activation functions and 8 dimensions per hidden layer. For context-sensitive term encodings (Fig. 3), a single layer was used, again with leaky-RELU as activation. For the density-based extension of ABEL-DRMM (Section 2.2.2), $l_w = 20$. For the confidence extension of ABEL-DRMM, $t_d = 100$, $t_c = 0.01$.

TERM-PACRR was also trained with Adam, with a learning rate of 0.001 and $\beta_1/\beta_2 = 0.9/0.999$ with batch size equal to 32. Following Hui et al. (2018), we used binary log-loss over pairs of a single positive and a single negative document of the

same query. Maximum query length l_q was set to 30 and maximum document length l_d was set to 300. Maximum kernel size $(l_g \times l_g)$ was set to (3×3) with 16 filters per size. Row-wise k-max pooling used $k = 2$. TERM-PACRR used a two-layer MLP with RELU activations and hidden layers with 7 dimensions to independently score each document-aware query-term encoding.

BCNN was trained using binary log-loss and AdaGrad (Duchi et al., 2011), with a learning rate of 0.08 and $L2$ regularization with $\lambda = 0.0004$. We used 50 convolution kernels (filters) of width $w = 4$ in each convolution layer, and two convolution/pooling blocks. Finally, batch sizes were set to 200. Snippets were truncated to 40 tokens. Questions were never truncated.

5.3 Official Submissions

We submitted 5 different systems to the BioASQ challenge, all of which consist of components described above.

- **AUEB-NLP-1**: Combo of 10 runs of TERM-PACRR for document retrieval (§2.1) followed by BCNN for snippet retrieval (§3).

- **AUEB-NLP-2**: Combo of 10 runs of ABEL-DRMM (§2.2) for document retrieval followed by BCNN for snippet retrieval.

- **AUEB-NLP-3**: Combo of 10 runs of TERM-PACRR and 10 runs of ABEL-DRMM followed by BCNN for snippet retrieval.

- **AUEB-NLP-4**: ABEL-DRMM with density extension (§2.2.2) for document retrieval followed by BCNN for snippet retrieval.

- **AUEB-NLP-5**: ABEL-DRMM with both density and confidence extensions (§2.2.2) for document retrieval followed by BCNN for

---

[14]We used NLTK v3.2.3. See https://www.nltk.org/api/nltk.tokenize.html.

[15]Available at http://participants-area.bioasq.org/results/6b/phaseA/. The names of our systems have been modified for the blind review.

snippet retrieval. This system was submitted for batches 2-5 only.

In combination (combo) systems, we obtained 10 versions of the corresponding model by retraining it 10 times with different random seeds, and we then used a simple voting scheme. If a document was ranked at position 1 by a model it got 10 votes, position 2 was 9 votes, until position 10 where it got 1 vote. Votes were then aggregated over all models in the combination. While voting did not improve upon the best single model, it made the results more stable across different runs.

5.4 Results

Results are given in Table 1. There are a number of things to note. First, for document retrieval, there is very little difference between our submitted models. Both PACRR- and DRMM-based models perform well (usually at the top or near the top) with less than 1 MAP point separating them. These systems were all competitive and for 4 of the 5 batches one was the top scoring system in the competition. On average the experimental ABEL-DRMM system (AUEB-NLP-4) scored best amongst AUEB submissions and in aggregate over all submissions, but by a small margin (0.1053 average MAP versus 0.1016 for TERM-PACRR). The exception was the high precision system (AUEB-NLP-5) which did worse in all metrics except F1, where it was easily the best system for the 4 batches it participated in. This is not particularly surprising, but impacted snippet selection, as we will see.

For snippet selection, all systems did well (AUEB-NLP-[1-4]) and it is hard to form a pattern that a base document retrieval model's results are more conducive to snippet selection. The exception is the high-precision document retrieval model of AUEB-NLP-5, which had by far the best scores for AUEB submissions and the challenge as a whole. The main reason for this is that the snippet retrieval component was trained assuming only relevant documents as input. Thus, if we fed it all 10 documents, even when some were not relevant, it could theoretically still rank a snippet from an irrelevant document high since it is not trained to combat this. By sending the snippet retrieval model only high precision document sets it focused on finding good snippets at the expense of potentially missing some relevant documents.

6   Related Work

Document ranking has been studied since the dawn of IR; classic term-weighting schemes were designed for this problem (Sparck Jones, 1972; Robertson and Sparck Jones, 1976). With the advent of statistical NLP and statistical IR, probabilistic language and topic modeling were explored (Zhai and Lafferty, 2001; Wei and Croft, 2006), followed recently by deep learning IR methods (Lu and Li, 2013; Hu et al., 2014; Palangi et al., 2016; Guo et al., 2016; Hui et al., 2017).

Most document relevance ranking methods fall within two categories: representation-based, e.g., Palangi et al. (2016), or interaction-based, e.g., Lu and Li (2013). In the former, representations of the query and document are generated independently. Interaction between the two only happens at the final stage, where a score is generated indicating relevance. End-to-end learning and backpropagation through the network tie the two representations together. In the interaction-based paradigm – which is where the models studied here fall – explicit encodings between pairs of queries and documents are induced. This allows direct modeling of exact- or near-matching terms (e.g., synonyms), which is crucial for relevance ranking. Indeed, Guo et al. (2016) showed that the interaction-based DRMM outperforms previous representation-based methods. On the other hand, interaction-based models are less efficient, since one cannot index a document representation independently of the query. This is less important, though, when relevance ranking methods rerank the top documents returned by a conventional IR engine, which is the scenario we consider here.

In terms of biomedical document and snippet retrieval, several methods have been proposed for BioASQ (Tsatsaronis et al., 2015), mostly based on traditional IR and ML techniques. For example, the system of Jin et al. (2017), which is the top scoring one for previous incarnations of BioASQ (UTSB team), uses an underlying graphical model for scoring coupled with a number of traditional IR techniques like pseudo-relevance feedback.

The most related work from the biomedical domain is that of Mohan et al. (2017), who use a deep learning architecture for document ranking. Like our systems they use interaction-based models to score and aggregate q-term matches relative to a document, however using different document-aware q-term representations – namely best match

DOCUMENT RETRIEVAL

System	F1	MAP	GMAP
Batch 1			
AUEB-NLP-1	0.2546	0.1246	0.0282
AUEB-NLP-2	0.2462	0.1229	0.0293
AUEB-NLP-*3*	0.2564	0.1271	0.0280
AUEB-NLP-4	0.2515	0.1255	0.0235
Top Competitor	0.2216	0.1058	0.0113
Batch 2			
AUEB-NLP-1	0.2264	0.1096	0.0148
AUEB-NLP-*2*	0.2473	0.1207	0.0200
AUEB-NLP-3	0.2364	0.1178	0.0161
AUEB-NLP-4	0.2350	0.1182	0.0161
AUEB-NLP-5	0.3609	0.1014	0.0112
Top Competitor	0.2265	0.1201	0.0183
Batch 3			
AUEB-NLP-1	0.2345	0.1122	0.0101
AUEB-NLP-2	0.2345	0.1147	0.0108
AUEB-NLP-3	0.2350	0.1135	0.0109
AUEB-NLP-4	0.2345	0.1137	0.0106
AUEB-NLP-5	0.4093	0.0973	0.0062
Top Competitor	0.2186	0.1281	0.0113
Batch 4			
AUEB-NLP-1	0.2136	0.0971	0.0070
AUEB-NLP-2	0.2148	0.0996	0.0069
AUEB-NLP-*3*	0.2134	0.1000	0.0068
AUEB-NLP-4	0.2094	0.0995	0.0064
AUEB-NLP-5	0.3509	0.0875	0.0044
Top Competitor	0.2044	0.0967	0.0073
Batch 5			
AUEB-NLP-1	0.1541	0.0646	0.0009
AUEB-NLP-2	0.1522	0.0678	0.0013
AUEB-NLP-3	0.1513	0.0663	0.0010
AUEB-NLP-*4*	0.1590	0.0695	0.0012
AUEB-NLP-5	0.1780	0.0594	0.0008
Top Competitor	0.1513	0.0680	0.0009

SNIPPET RETRIEVAL

System	F1	MAP	GMAP
Batch 1			
AUEB-NLP-1	0.1296	0.0687	0.0029
AUEB-NLP-2	0.1347	0.0665	0.0026
AUEB-NLP-3	0.1329	0.0661	0.0028
AUEB-NLP-*4*	0.1297	0.0694	0.0024
Top Competitor	0.1028	0.0710	0.0002
Batch 2			
AUEB-NLP-1	0.1329	0.0717	0.0034
AUEB-NLP-2	0.1434	0.0750	0.0044
AUEB-NLP-3	0.1355	0.0734	0.0033
AUEB-NLP-4	0.1397	0.0713	0.0037
AUEB-NLP-*5*	0.1939	0.1368	0.0045
Top Competitor	0.1416	0.0938	0.0011
Batch 3			
AUEB-NLP-1	0.1563	0.1331	0.0046
AUEB-NLP-2	0.1494	0.1262	0.0034
AUEB-NLP-3	0.1526	0.1294	0.0038
AUEB-NLP-4	0.1519	0.1293	0.0038
AUEB-NLP-*5*	0.2744	0.2314	0.0068
Top Competitor	0.1877	0.1344	0.0014
Batch 4			
AUEB-NLP-1	0.1211	0.0716	0.0009
AUEB-NLP-2	0.1307	0.0821	0.0011
AUEB-NLP-3	0.1251	0.0747	0.0009
AUEB-NLP-4	0.1180	0.0750	0.0009
AUEB-NLP-*5*	0.1940	0.1425	0.0017
Top Competitor	0.1306	0.0980	0.0006
Batch 5			
AUEB-NLP-1	0.0768	0.0357	0.0003
AUEB-NLP-2	0.0728	0.0405	0.0004
AUEB-NLP-3	0.0747	0.0377	0.0004
AUEB-NLP-4	0.0790	0.0403	0.0004
AUEB-NLP-*5*	0.0778	0.0526	0.0003
Top Competitor	0.0542	0.0475	0.0001

Table 1: Performance on BioASQ Task 6b, Phase A (batches 1–5) for document and snippet retrieval (left and right tables, respectively). Systems described in Section 5.3. The italicised system is the top scoring system from AUEB's entries and if also in bold, is the top from all official entries in that batch. *Top* is by MAP, the official metric of BioASQ. *Top Competitor* is the top scoring entry – by MAP– that is not among AUEB's submissions.

d-term distance scores. Also unlike our work, they focus on user click data as a supervised signal, and they use context-insensitive representations of document-query term interactions.

There are several studies on deep learning systems for snippet selection which aim to improve the classification and ranking of snippets extracted from a document based on a specific query. Wang and Nyberg (2015) use a stacked bidirectional LSTM (BILSTM); their system gets as input a question and a sentence, it concatenates them in a single string and then forwards that string to the input layer of the BILSTM. Rao et al. (2016) employ a neural architecture to produce representations of pairs of the form (*question, sentence*) and to learn to rank pairs of the form (*question, relevant sentence*) higher than pairs of the form (*question, irrelevant sentence*) using Noise-Contrastive Estimation. Finally, Amiri et al. (2016) use autoen-coders to learn to encode input texts and use the resulting encodings to compute similarity between text pairs. This is similar in nature to BCNN, the main difference being the encoding mechanism.

7 Conclusions

We presented the models, experimental set-up, and results of AUEB's submissions to the document and snippet retrieval tasks of the sixth year of the BioASQ challenge. Our results show that deep learning models are not only competitive in both tasks, but in aggregate were the top scoring systems. This is in contrast to previous years where traditional IR systems tended to dominate. In future years, as deep ranking models improve and training data sets get larger, we expect to see bigger gains from deep learning models.

References

Hadi Amiri, Philip Resnik, Jordan Boyd-Graber, and Hal Daumé III. 2016. Learning text pair similarity with context-sensitive autoencoders. In *Proceedings of the 54th Annual Meeting of the Association for Computational Linguistics (Volume 1: Long Papers)*, pages 1882–1892, Berlin, Germany.

Shaojie Bai, J Zico Kolter, and Vladlen Koltun. 2018. An empirical evaluation of generic convolutional and recurrent networks for sequence modeling. *arXiv preprint arXiv:1803.01271*.

John Duchi, Elad Hazan, and Yoram Singer. 2011. Adaptive subgradient methods for online learning and stochastic optimization. *Journal of Machine Learning Research*, 12:2121–2159.

Jiafeng Guo, Yixing Fan, Qingyao Ai, and W. Bruce Croft. 2016. A deep relevance matching model for ad-hoc retrieval. In *Proceedings of the 25th ACM International Conference on Information and Knowledge Management*, pages 55–64, Indianapolis, IN.

Baotian Hu, Zhengdong Lu, Hang Li, and Qingcai Chen. 2014. Convolutional neural network architectures for matching natural language sentences. In *Advances in Neural Information Processing Systems 27*, pages 2042–2050.

Kai Hui, Andrew Yates, Klaus Berberich, and Gerard de Melo. 2017. PACRR: A position-aware neural IR model for relevance matching. In *Proceedings of the Conference on Empirical Methods in Natural Language Processing*, pages 1049–1058, Copenhagen, Denmark.

Kai Hui, Andrew Yates, Klaus Berberich, and Gerard de Melo. 2018. Co-PACRR: A context-aware neural IR model for ad-hoc retrieval. In *Proceedings of the 11th ACM International Conference on Web Search and Data Mining*, pages 279–287, Marina Del Rey, CA.

Zan-Xia Jin, Bo-Wen Zhang, Fan Fang, Le-Le Zhang, and Xu-Cheng Yin. 2017. A multi-strategy query processing approach for biomedical question answering: Ustb_prir at BioASQ 2017 Task 5B. In *BioNLP 2017*, pages 373–380.

Diederik P Kingma and Jimmy Ba. 2014. Adam: A method for stochastic optimization. *arXiv preprint arXiv:1412.6980*.

Robert Krovetz. 1993. Viewing morphology as an inference process. In *Proceedings of the 16th Annual International ACM SIGIR Conference on Research and Development in Information Retrieval*, pages 191–202, Pittsburgh, PA.

Zhengdong Lu and Hang Li. 2013. A deep architecture for matching short texts. In *Proceedings of the 26th International Conference on Neural Information Processing Systems - Volume 1*, pages 1367–1375, Lake Tahoe, NV.

Ryan McDonald, Georgios-Ioannis Brokos, and Ion Androutsopoulos. 2018. Deep relevance ranking using enhanced document-query interactions. In *Proceedings of the Conference on Empirical Methods in Natural Language Processing*, Brussels, Belgium.

Tomas Mikolov, Ilya Sutskever, Kai Chen, Greg S Corrado, and Jeff Dean. 2013. Distributed representations of words and phrases and their compositionality. In *Proceedings of the 26th International Conference on Neural Information Processing Systems - Volume 2*, pages 3111–3119, Lake Tahoe, Nevada.

Sunil Mohan, Nicolas Fiorini, Sun Kim, and Zhiyong Lu. 2017. Deep learning for biomedical information retrieval: Learning textual relevance from click logs. In *BioNLP 2017*, pages 222–231.

Hamid Palangi, Li Deng, Yelong Shen, Jianfeng Gao, Xiaodong He, Jianshu Chen, Xinying Song, and Rabab Ward. 2016. Deep sentence embedding using long short-term memory networks: Analysis and application to information retrieval. *IEEE/ACM Transactions on Audio, Speech and Language Processing*, 24(4):694–707.

Jinfeng Rao, Hua He, and Jimmy J. Lin. 2016. Noise-contrastive estimation for answer selection with deep neural networks. In *CIKM*, pages 1913–1916, Indianapolis, IN.

Stephen Robertson, S. Walker, S. Jones, M. M. Hancock-Beaulieu, and M. Gatford. 1995. Okapi at TREC3. In *Overview of the Third Text Retrieval Conference*, pages 109–126.

Stephen E. Robertson and Karen Sparck Jones. 1976. Relevance weighting of search terms. *Journal of the Association for Information Science and Technology*, 27(3):129–146.

Aliaksei Severyn and Alessandro Moschitti. 2015. Learning to rank short text pairs with convolutional deep neural networks. In *Proceedings of the 38th International ACM SIGIR Conference on Research and Development in Information Retrieval*, pages 373–382, Santiago, Chile.

Karen Sparck Jones. 1972. A statistical interpretation of term specificity and its application in retrieval. *Journal of documentation*, 28(1):11–21.

George Tsatsaronis, Georgios Balikas, Prodromos Malakasiotis, Ioannis Partalas, Matthias Zschunke, Michael R. Alvers, Dirk Weissenborn, Anastasia Krithara, Sergios Petridis, Dimitris Polychronopoulos, Yannis Almirantis, John Pavlopoulos, Nicolas Baskiotis, Patrick Gallinari, Thierry Artiéres, Axel-Cyrille Ngonga Ngomo, Norman Heino, Eric Gaussier, Liliana Barrio-Alvers, Michael Schroeder, Ion Androutsopoulos, and Georgios Paliouras. 2015. An overview of the BioASQ large-scale biomedical semantic indexing and question answering competition. *BMC Bioinformatics*, 16(1):138.

Di Wang and Eric Nyberg. 2015. A long short-term memory model for answer sentence selection in question answering. In *Proceedings of the 53rd Annual Meeting of the Association for Computational Linguistics and the 7th International Joint Conference on Natural Language Processing (Volume 2: Short Papers)*, pages 707–712, Beijing, China.

Xing Wei and W Bruce Croft. 2006. LDA-based document models for ad-hoc retrieval. In *Proceedings of the 29th Annual International ACM SIGIR Conference on Research and Development in Information Retrieval*, pages 178–185, Seattle, WA.

Wenpeng Yin, Hinrich Schütze, Bing Xiang, and Bowen Zhou. 2016. ABCNN: Attention-based convolutional neural network for modeling sentence pairs. *Transactions of the Association for Computational Linguistics*, 4:259–272.

Chengxiang Zhai and John Lafferty. 2001. A study of smoothing methods for language models applied to ad hoc information retrieval. In *Proceedings of the 24th Annual International ACM SIGIR Conference on Research and Development in Information Retrieval*, pages 334–342, New Orleans, LA.

MindLab Neural Network Approach at BioASQ 6B

Andrés Rosso-Mateus, Fabio A. González
MindLab Research Group
Universidad Nacional de Colombia
Bogotá, Colombia
`aerossom,fagonzalezo@unal.edu.co`

Manuel Montes-y-Gómez
Laboratorio de Tecnologías del Lenguaje
INAOE
Puebla, Mexico
`mmontesg@inaoep.mx`

Abstract

Biomedical Question Answering is concerned with the development of methods and systems that automatically find answers to natural language posed questions. In this work, we describe the system used in the BioASQ Challenge task 6b for document retrieval and snippet retrieval (with particular emphasis in this subtask). The proposed model makes use of semantic similarity patterns that are evaluated and measured by a convolutional neural network architecture. Subsequently, the snippet ranking performance is improved with a pseudo-relevance feedback approach in a later step. Based on the preliminary results, we reached the second position in snippet retrieval sub-task.

1 Introduction

The development of methods that contribute to bypass the manual checking of candidate documents, is playing an important role in the closed domain information access and will be the next step in information retrieval systems (Zadeh, 2006). The number of published documents grows continuously and pertain to a large variety of topics. More than 3000 articles are indexed every day in biomedical journals (Tsatsaronis et al., 2012), making it harder for patients and physicians to access valuable information. The produced data needs to be mined in order to have a positive impact on public health, although it also represents a challenge.

The Question Answering (QA) paradigm can help to retrieve concise information in a natural way, given the precise answer and the supporting passages for any information need. The research in QA has been pulled by organizations and challenges that encourage academic community to develop new systems and methods to tackle this complex task.

One of the most important challenges is BioASQ, focused on indexing and question answering tasks over biomedical articles (Tsatsaronis et al., 2015).

In this work, we describe our first participation in the sixth edition of the BioASQ challenge. We participated in task B, which is composed of two phases.

- Phase A: Given a question the system must return relevant concepts (from designated terminologies and ontologies), relevant documents (from PubMed articles baseline (pub)), relevant snippets (extracted from articles), and relevant RDF triples (from designated ontologies) (Tsatsaronis et al., 2015).

- Phase B: Given a question and a set of relevant articles and snippets. The system must provide an exact answer (e.g., named entities) and ideal answers (summaries) (Tsatsaronis et al., 2015).

BioASQ challenge rules allow teams to participate in any of the two phases, and also send results for any or all of the sub-tasks in the desired phase. We chose **Phase A** for our first participation, and we submitted results for (1) document retrieval and (2) snippet retrieval.

2 Methods

2.1 Model Architecture

The whole system is composed of two main modules as shown in Figure 1. A document retrieval module searches the PubMed Baseline Repository (MBR) (pub) for relevant articles, and a fine-grained information retrieval model to identify the 10 most relevant snippets. For document retrieval we used Elastic Search (ES) engine (Gormley and Tong, 2015) with BM25 as relevance ranking function (Agichtein et al., 2006). To improve

40

Proceedings of the 2018 EMNLP Workshop BioASQ: Large-scale Biomedical Semantic Indexing and Question Answering, pages 40–46
Brussels, Belgium, November 1st, 2018. ©2018 Association for Computational Linguistics

the performance we added to the index the title, abstract and concepts for all the documents. When a search is performed, all fields are compared against the search query.

Most related documents are analyzed in depth. We split the documents into sentences and those sentences feed the snippet retrieval stage. We process the snippets with a Convolutional Neural Network (CNN) to obtain a semantic similarity relevance score.

Finally, the scored snippets are sorted in descending order and the 10 with the highest scores are selected. The documents are re-ranked based on a standardized linear combination between Elastic Search score and the average of their snippets scores. The 10 most related documents and snippets were submitted to BioASQ server.

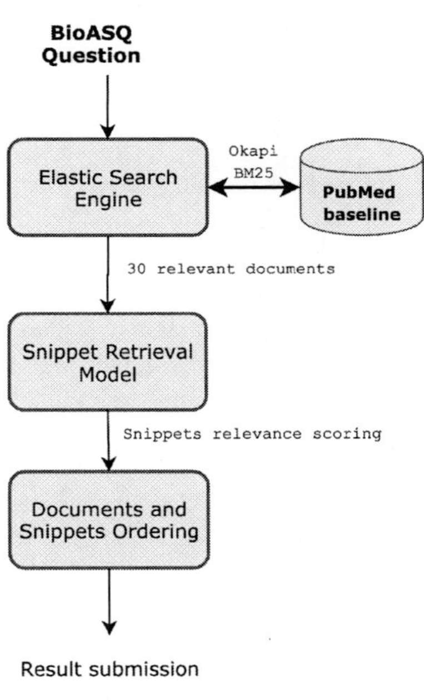

Figure 1: BioASQ Model Diagram

A detailed description of the model will be presented in the following sections.

2.2 Document Retrieval

Question answering systems make use of document retrieval methods to provide relevant documents that could contain the answer to a user's question.

The document retrieval system affects the question answering method effectiveness: if a retrieval system does not find relevant documents for a question, the later stages will inevitably fail.

In the BioASQ challenge the document retrieval system has to index approximately 27 millions medical articles. This huge amount of data makes it necessary to have a high-performance platform. We used Elastic Search (ES), a standalone search engine written in Java that stores the related data in a sophisticated format optimized for language based query searches (Gormley and Tong, 2015). ES is also easily scalable and comes with a default configuration that makes the whole learning process easy.

Elastic Search uses by default BM25, which is an improvement of TF-IDF ranking function that takes into account the length of documents and queries.

2.3 Snippet Retrieval

The main assumption of the snippet retrieval model is that the question and the answer are semantically related based on their terms. So the question-answer inter-correlation is given by the relationship between their component terms.

The proposed method has two stages. The first one (training phase) has the objective to learn the similarity patterns between question-answer pairs. In the second stage (prediction and re-ordering) the similarity model is used to obtain the first ranking between question and answer pairs, then a reordering is carried out using pseudo-relevance feedback based on the terms from the most related answer in the first ordering. The whole process is depicted in Figure 2.

The training phase is carried out to obtain the similarity model, then this model is used in the testing phase to rank the question-answer pairs. During training: (1) question-answer pairs (QA-pairs) are pre-processed, (2) the similarity matrix between QA-pairs terms is calculated, and (3) a convolutional neural network model is trained to predict the relevance of the answer to the question. Once the model is built it can be used to predict the rank of candidate answers. At testing time, for a particular question, the model is applied to predict the relevance score of the set of candidate answers, (4) answers are ranked according to their scores, (5) answers are re-ranked according to their similarity with the highest ranked answer at step (4), producing a new ranking of the answers.

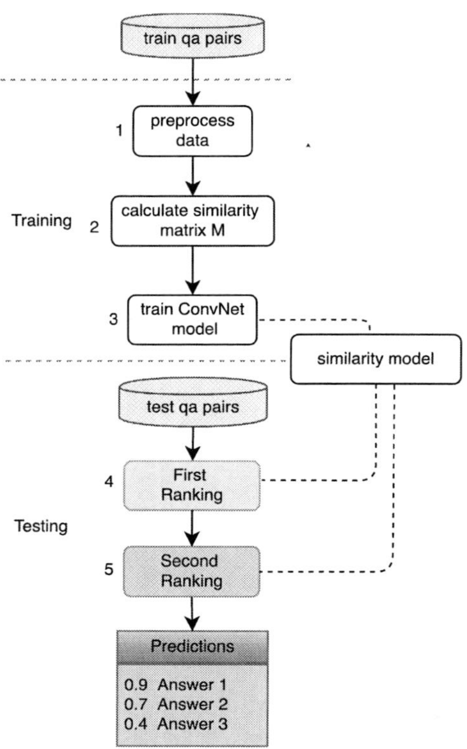

Figure 2: Two-Step Similarity Scoring Model Architecture

2.3.1 Step 1. Preprocess Data

Questions and candidate answers are processed using: tokenization to delimit terms; lowercasing to standardize the terms; POS-tagging, using the NLTK POS-tagger (Bird, 2006), to extract syntactical information that will be used in salience weighting; and transforming terms to a word2vec vector representation (Mikolov et al., 2013), to make possible their semantic similarity comparison.

2.3.2 Step 2. Calculate Similarity Matrix

The similarity matrix M represents the semantic relatedness of the i-th question term and the j-th answer term according to a similarity measure. Each element $M_{i,j}$ of this matrix is a composition of a similarity score and a salience score as described by Eq. 1.

$$M_{i,j} = scos(q_i, a_j) * sal(q_i, a_j) \qquad (1)$$

2.3.3 Similarity Score

The similarity score for question-answer pair terms (q_i, a_j) is calculated using cosine similarity between their word2vec vectors as indicated by Eq. 2.

$$scos(q_i, a_j) = 0.5 + \frac{q_i \cdot a_j}{2 \, \|q_i\|_2 \, \|a_j\|_2} \qquad (2)$$

In the case that there does not exist the word2vec representation for one of the terms, their similarity is measured based on their distance in Wordnet. In particular, we use as similarity measure the edge distance between the first common concept related with q_i and a_j (Wu and Palmer, 1994). If there is not a common concept between the terms, then we calculate the Levenshtein distance between the words (Levenshtein, 1966), defined as the number of operations (insertions and eliminations of characters) needed to transform q_i to a_j.

2.3.4 Salience Weighting

As not all terms are equally informative for measuring text similarities (Liu et al., 2009; Dong et al., 2015), we consider weighting the terms from the question and the answer based on part of speech functions: verbs, nouns, and adjectives are considered to be the most relevant. We model this information through a salience score.

The salience score is calculated as follows. If both terms are relevant then their score is 1. If only one of the terms is important then the score is 0.6, in the case none of them is relevant the score is 0.3. The salience function is defined in the Eq. 3.

$$sal(q_i, a_j) = \begin{cases} 1 & if\ imp(q_i) + imp(a_j) = 2 \\ 0.6 & if\ imp(q_i) + imp(a_j) = 1 \\ 0.3 & if\ imp(q_i) + imp(a_j) = 0 \end{cases}$$

$$(3)$$

Where $imp(q_i)$ and $imp(a_j)$ are the evaluation of importance weighting function for every question and answer term. The related function returns 1 if the term is a verb, noun or adjective, otherwise, returns 0.

Finally, we sort the calculated matrix M leaving the most related terms in the top left cell, and if the number of rows or columns exceeds 40, the remaining data is truncated. This step provides an invariable representation of the similarity patterns that can be exploited by the convolutional network.

2.3.5 Step 3. Convolutional Model

Convolutional neural networks (CNN) are a popular method for image analysis thanks to their ability to capture spatial invariant patterns. In the proposed method, they play a similar role, but instead of receiving an input image the CNN receives the similarity matrix M. The hypothesis is that it will be able to identify term-similarity patterns that help to determine the relevance of a question-answer pair. Patterns identified by the CNN are sub-sampled by a pooling layer. The output of the pooling layer feeds a fully-connected layer. Finally, the output of the model is generated by a sigmoid unit. This output corresponds to a score, **simScore**(q, a), that can be interpreted as a degree of relatedness between the question q and the answer a.

The architecture of the convolutional model is depicted in Figure 3.

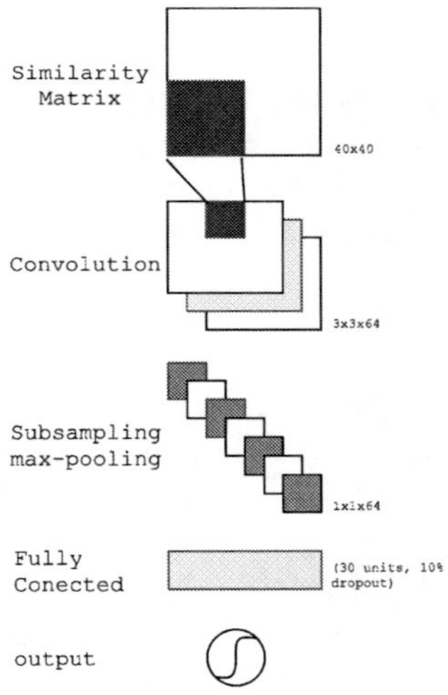

Figure 3: Convolutional Neural Network Model

2.3.6 Step 4 and 5. Two Ranking Stages

During the testing phase, a new query along with candidate answers are presented to the method. The candidate answers $(a_1, a_2, ..., a_k)$ are ranked using the CNN model producing the first rank of them. Based on the premise that the first candidate answer, a^*, is expected to be highly correlated with the question q, a second score, $simScore(a^*, a_k)$, is calculated by comparing each candidate answer with the highest ranked answer. A new ranking is calculated by using a new score corresponding to a linear combination of the first and second score as is shown in Eq. 4.

$$\begin{aligned} finalScore(q, a_k) = \\ (1 - \alpha) * simScore(q, a_k) + \\ \alpha * simScore(a^*, a_k) \end{aligned} \quad (4)$$

As we are introducing a weighting term α to scale the second score, we calculated this term based on the exploration carried out in a validation partition, which gives 0.32 as the optimal value.

This strategy promotes candidate answers which share similar terms with the highest ranked answer. This is a strategy analogous to pseudo-relevance feedback in information retrieval (Riezler et al., 2007), where the original query is extended with terms from the highest ranked documents.

2.4 Experiments

We indexed the full data of 2017 PubMed baseline in ElasticSearch engine (ES) version 6.2.2 with the default configuration. The number of processed files were 928 and the total number of medical articles was 26,759,399. For each article, we extracted the title, MESH concepts and abstract to be indexed. The indexing time was around 18 hours in an Intel Xeon processor Intel(R) at 2.60GHz with 82 GB RAM and GeForce GTX TITAN X.

The training was done with the question and answer pairs from 2016, 2017 and 2018 BioASQ Task B training data-set. The total number of question-answer pairs used were 124,144. The obtained data-set was very unbalanced, only 18% of the total number of pairs are labeled as an answer. To balance the data-set, the sample extraction in training phase is done with the same number of positives and negative samples, this strategy is also applied in the validation phase.

The model training was done using RMSprop optimization algorithm with 256 samples in mini-batch and the defined loss function is binary cross entropy. The number of maximum epochs was set to 500. In each epoch, we evaluate MAP and MRR, and after 20 epochs without any improvement in MAP metric, we apply early stopping to avoid over-fitting.

2.4.1 Model parameters

The model hyper-parameters were tuned using hyper-parameter exploration. The parameters chosen are listed next.

- **Convolution Parameters:** The number of convolutional filters used are 64, width 3 and length 3, the stride used is 1 without padding.

- **Convolution Activation Function:** After a convolutional layer, it is useful to apply a nonlinear layer (Goodfellow et al., 2016). We tested different activation functions and RELU gave us the best performance.

- **Pooling Layers:** For the pooling layer, we used max pooling.

- **Dropout Layer:** We add a dropout layer as a regularization strategy (Srivastava et al., 2014), setting the parameter in 10%.

Finally, the number of parameters to learn in our model is not very high (3,198) compared with other Convolution Neural approaches used in similar tasks (Question Answering) which are in order of millions and hundreds of thousands (Severyn and Moschitti, 2015; He and Lin, 2016)

2.5 Model Tuning

In this section, we will describe the strategy to improve the overall performance of our system. The metrics were calculated over the training dataset released by BioASQ for the 6th version.

- Mesh concept indexing: Document retrieval is mainly based on Elastic Search keyword matching evaluation with BM25 ranking function. We used a cross-fields query approach which looks for each term in the title, abstract and concepts indexed fields. Considering the retrieval of 10 most related documents, the performance using cross-fields approach were (Recall = 0.24, MAP = 0.19) while not using this were (Recall=0.278, MAP= 0.221).

- Word representation: The choice of a good word representation is important to generate a semantically good model where relations between terms or sentences are more easy to establish. We tested our system using different pre-trained word2vec models and the best representation was the skip-gram

model provided by NLPLab, which is trained on Wikipedia and PubMed abstracts (Moen and Ananiadou, 2013). The MAP score in the snippet retrieval sub-task improved from 0.126 to 0.142.

- Training dataset generation: The training corpus was generated with questions and answer passages extracted from 2016, 2017 and 2018 BioASQ training datasets. We tested different rates of negative samples (passages in related documents that does contain the answer) in order to increase the negative sample coverage. This assumption is based on the hypothesis that it is not easy to determine that a related snippet does not contain the answer. With a higher negative sample generation, these cases are more common, and the method can learn a better discriminant function. The rate that experimentally achieved the best results considers using 10 negative samples per 1 positive sample. The MAP score in snippet retrieval sub-task, improved using 6b training partition from 0.142 to 0.151.

- Document re-ranking: After obtaining the similarity scores for snippets and the initial Elastic Search BM25 score for documents, the scores are combined as follows, eq. 5.

$$doc\_score(q, d_k) =$$
$$(1 - \alpha) * es\_score(q, d_k) +$$
$$\alpha * avg(sim\_score(q, doc_k\_snippet\_j)) \tag{5}$$

where, $avg(sim\_score(q, doc_k\_snippet\_j))$ is the averaged similarity score between snippets of $document\_k$ and the query q. The calculated score is used to return the final list of documents. The parameter for the linear combination α, is calculated in evaluation step ($\alpha = 0.09$). Experimental results show that the contribution of Elastic Search score is higher (0.91). The improvement in document retrieval metrics was not significant but was around a 0.1 point in MAP and RECALL.

3 Results and Discussion

In this section, we present the preliminary results for the sixth version of BioASQ challenge in task

B phase A. The first sub-task is to retrieve the most related articles based on a question posed in natural language. The second one is to retrieve the snippets that have more correlation with the question in order to use them to compose an answer. The answer composition is carried out in phase B, which was not the scope of our participation.

3.1 Document Retrieval

The results shown in the Table 1 reveal, that our ES document retrieval implementation did not have a good performance, the recall obtained is low in all the batches. In the first batch, we had a technical issue that corrupted the results, it also happened for snippet retrieval. The best result was obtained in batch 3 (Recall = 0.49), the team leader in this batch reached 0.56, an important difference. As it was mentioned before, document retrieval is very important for snippet retrieval, it is the first information filter and it feeds the method to rank their snippets. Despite the low recall in this step, we will see in the next section that snippet retrieval scores are very promising.

Batch	Document Retrieval	
	Mean precision	Recall
	F-Measure	MAP
1	-	-
	-	-
2	0.1150	0.4685
	0.1621	0.0709
3	0.1320	0.4984
	0.1782	0.0891
4	0.1240	0.4467
	0.1717	0.0846
5	0.0890	0.2961
	0.1260	0.0540

Table 1: Document retrieval results

3.2 Snippet Retrieval

In this stage, we analyzed in depth the returned set of documents from the previous method, and identify the text snippets that can answer the posed question.

Based on the evidence shown in Table 2, the snippet retrieval approach obtained a good performance. We could have had a better performance in snippet retrieval with a higher score in document retrieval, but it was enough to reach the second position in all the batches except the first one (due to the technical issue).

We can state that the proposed method exhibits a very competitive performance compared with other methods.

Batch	Snippet Retrieval	
	Mean precision	Recall
	F-Measure	MAP
1	-	-
	-	-
2	0.1111	0.2426
	0.1416	0.0938
3	0.1614	0.2657
	0.1877	0.1344
4	0.1043	0.2180
	0.1306	0.0980
5	0.0404	0.1134
	0.0542	0.0475

Table 2: Snippet retrieval results

4 Conclusion

This work presents our first participation in BioASQ (task B phase A) document retrieval and snippet retrieval tasks. Our system was based on Elastic Search platform with the BM25 scoring function for document retrieval. For snippet retrieval, we presented a novel method based on a convolutional neural network with a pseudo-relevance-feedback re-ranking step.

The preliminary results are promising in snippets retrieval sub-task, where the proposed method reached the second position in all batches except the first one. This result gain in importance based on the fact that the chosen approach for document retrieval sub-task did not give good results.

The future work will be focused on improving document retrieval sub-task to feed the snippet retrieval method with a more complete (and higher quality) list of candidate answers. We will also work in a better question-answer pair representation with the incorporation of structured data sources for gain information.

Acknowledgments

COLCIENCIAS, REF. Agreement N°. 727, 2016 provided financial as well as logistical and planning support. Mindlab research group (Universidad Nacional de Colombia sede Bogotá) with the coperation of INAOE (Instituto Nacional de Astrofísica, Óptica y Electrónica) also provided technical support for this work.

References

Pubmed baseline repository.

Eugene Agichtein, Eric Brill, and Susan Dumais. 2006. Improving web search ranking by incorporating user behavior information. In *Proceedings of the 29th annual international ACM SIGIR conference on Research and development in information retrieval*, pages 19–26. ACM.

Steven Bird. 2006. Nltk: the natural language toolkit. In *Proceedings of the COLING/ACL on Interactive presentation sessions*, volume 1, pages 69–72. ACL.

Li Dong, Furu Wei, Ming Zhou, and Ke Xu. 2015. Question answering over freebase with multi-column convolutional neural networks. In *ACL*, volume 1, pages 260–269.

Ian Goodfellow, Yoshua Bengio, and Aaron Courville. 2016. *Deep learning*. MIT Press.

Clinton Gormley and Zachary Tong. 2015. *Elasticsearch: The Definitive Guide: A Distributed Real-Time Search and Analytics Engine*. " O'Reilly Media, Inc.".

Hua He and Jimmy J Lin. 2016. Pairwise word interaction modeling with deep neural networks for semantic similarity measurement. In *HLT-NAACL*, volume 1, pages 937–948.

Vladimir I Levenshtein. 1966. Binary codes capable of correcting deletions, insertions, and reversals. In *Soviet physics doklady*, volume 10, pages 707–710.

Feifan Liu, Deana Pennell, Fei Liu, and Yang Liu. 2009. Unsupervised approaches for automatic keyword extraction using meeting transcripts. In *Proceedings of human language technologies.*, volume 1, pages 620–628. ACL.

Tomas Mikolov, Kai Chen, Greg Corrado, and Jeffrey Dean. 2013. Efficient estimation of word representations in vector space. *arXiv preprint arXiv:1301.3781*.

SPFGH Moen and Tapio Salakoski2 Sophia Ananiadou. 2013. Distributional semantics resources for biomedical text processing. In *Proceedings of the 5th International Symposium on Languages in Biology and Medicine, Tokyo, Japan*, pages 39–43.

Stefan Riezler, Alexander Vasserman, Ioannis Tsochantaridis, Vibhu Mittal, and Yi Liu. 2007. Statistical machine translation for query expansion in answer retrieval. In *ACL*.

Aliaksei Severyn and Alessandro Moschitti. 2015. Learning to rank short text pairs with convolutional deep neural networks. In *38th ACM SIGIR*.

Nitish Srivastava, Geoffrey E Hinton, Alex Krizhevsky, Ilya Sutskever, and Ruslan Salakhutdinov. 2014. Dropout: a simple way to prevent neural networks from overfitting. *Journal of Machine Learning Research*, 15(1):1929–1958.

George Tsatsaronis, Georgios Balikas, Prodromos Malakasiotis, Ioannis Partalas, Matthias Zschunke, Michael R Alvers, Dirk Weissenborn, Anastasia Krithara, Sergios Petridis, Dimitris Polychronopoulos, Yannis Almirantis, John Pavlopoulos, Nicolas Baskiotis, Patrick Gallinari, Thierry Artiéres, Axel-Cyrille Ngonga Ngomo, Norman Heino, Eric Gaussier, Liliana Barrio-Alvers, Michael Schroeder, Ion Androutsopoulos, and Georgios Paliouras. 2015. An overview of the BIOASQ large-scale biomedical semantic indexing and question answering competition. *BMC Bioinformatics*, 16(1):138.

George Tsatsaronis, Michael Schroeder, Georgios Paliouras, Yannis Almirantis, Ion Androutsopoulos, Eric Gaussier, Patrick Gallinari, Thierry Artieres, Michael R Alvers, Matthias Zschunke, et al. 2012. Bioasq: A challenge on large-scale biomedical semantic indexing and question answering. In *AAAI fall symposium: Information retrieval and knowledge discovery in biomedical text*.

Zhibiao Wu and Martha Palmer. 1994. Verbs semantics and lexical selection. In *32nd Proceedings ACL*, volume 1, pages 133–138. Association for Computational Linguistics.

Lotfi A Zadeh. 2006. From search engines to question answering systems—the problems of world knowledge, relevance, deduction and precisiation. *Capturing Intelligence*, 1:163–210.

AttentionMeSH: Simple, Effective and Interpretable Automatic MeSH Indexer

Qiao Jin*
University of Pittsburgh
qiao.jin@pitt.edu

Bhuwan Dhingra*
Carnegie Mellon University
bdhingra@cs.cmu.edu

William W. Cohen†
Google, Inc.
wcohen@google.com

Xinghua Lu
University of Pittsburgh
xinghua@pitt.edu

Abstract

There are millions of articles in PubMed database. To facilitate information retrieval, curators in the National Library of Medicine (NLM) assign a set of Medical Subject Headings (MeSH) to each article. MeSH is a hierarchically-organized vocabulary, containing about 28K different concepts, covering the fields from clinical medicine to information sciences. Several automatic MeSH indexing models have been developed to improve the time-consuming and financially expensive manual annotation, including the NLM official tool – Medical Text Indexer, and the winner of BioASQ Task5a challenge – DeepMeSH. However, these models are complex and not interpretable. We propose a novel end-to-end model, AttentionMeSH, which utilizes deep learning and attention mechanism to index MeSH terms to biomedical text. The attention mechanism enables the model to associate textual evidence with annotations, thus providing interpretability at the word level. The model also uses a novel masking mechanism to enhance accuracy and speed. In the final week of BioASQ Chanllenge Task6a, we ranked 2nd by average MiF using an on-construction model. After the contest, we achieve close to state-of-the-art MiF performance of ~ 0.684 using our final model. Human evaluations show AttentionMeSH also provides high level of interpretability, retrieving about 90% of all expert-labeled relevant words given an MeSH-article pair at 20 output.

1 Introduction

MEDLINE is a database containing more than 24 million biomedical journal citations by 2018[1].

PubMed provides free access to MEDLINE for worldwide researchers. To facilitate information storage and retrieval, curators at the National Library of Medicine (NLM) assign a set of Medical Subject Headings (MeSH) to each article. MeSH[2] is a hierarchically-organized terminology developed by NLM for indexing and cataloging biomedical texts like MEDLINE articles. MeSH has about 28 thousand terms by 2018[3], covering the fields from clinical medicine to information sciences. Indexers examine the full article and annotate it with MeSH terms according to rules set by NLM[4]. Its estimated that indexing an article costs $9.4 on average (Mork et al., 2013), and there are more than 813,500 citations added to MEDLINE in 2017[5]. Indexing all citations manually would cost several million dollars in one year. Thus, several automatic annotation models have been developed to improve the time-consuming and financially expensive manual annotation. We will discuss these models in section 2.1.

Automatical annotating PubMed abstracts with MeSH terms is hard in several aspects: There are 28 thousand possible classes and even more of their combinations. The frequencies of different MeSH terms also vary a lot: The most frequent MeSH term is 'Humans' and it is annotated to more than 8 million articles in the MEDLINE database; while the 20,000th frequent MeSH 'Hypnosis, Anesthetic' is indexed to only about 200 articles (Peng et al., 2016). It causes severe class imbalance problems. Above difficulties are further complicated by the fact that indexers at the NLM usually inspect the whole articles to

---

*These authors contribute equally to the paper.
†This work was done while the author was at CMU.
[1]https://www.nlm.nih.gov/pubs/
factsheets/medline.html

[2]https://www.nlm.nih.gov/mesh
[3]https://www.nlm.nih.gov/pubs/
factsheets/mesh.html
[4]https://www.nlm.nih.gov/bsd/indexing/
training/TIP_010.html
[5]https://www.nlm.nih.gov/bsd/bsd_key.
html

Proceedings of the 2018 EMNLP Workshop BioASQ: Large-scale Biomedical Semantic Indexing and Question Answering, pages 47–56
Brussels, Belgium, November 1st, 2018. ©2018 Association for Computational Linguistics

do the annotation, but the challenge only provides PubMed abstracts and titles, which might not be enough to find all MeSH terms. We will discuss it more detailedly in section 5.

Deep learning is a subtype of machine learning that arranges the computational models in multiple processing layers to learn the representations of data with multiple levels of abstractions as well as the mapping from these features to the output (LeCun et al., 2015). Attention is a strategy for deep learning models to learn both the mapping from input to output and the relevance between input parts and output parts (Bahdanau et al., 2014). The learnt relevance helps improve the mapping performance as well as provide interpretability. We will discuss relevant works of deep learning in automatic annotations in section 2.2.

Here we propose a novel model, AttentionMeSH, which utilizes deep learning and attention mechanism to index MeSH terms to biomedical texts and provides interpretation at the word level. Each abstract, together with title and journal name, is tokenized to words, then the model feeds word vectors to a bidirectional gated recurrent unit (BiGRU) to derive word representations with contextual information (Schuster and Paliwal, 1997; Cho et al., 2014). We narrow down the MeSH term vocabulary for each abstract using a masking mechanism. Then for each candidate MeSH term, the model calculates the attention attribution over words. Next, each MeSH term gets a specific document representations by MeSH-specific attention-weighted sum of the word vectors. Finally, the model uses nonlinear layers to classify each MeSH term using the learnt MeSH-specific document representation.

We participated in BioASQ Challenge Task6A while developing the model. We achieve close to state-of-the-art performance with an on-construction model in the final week of the contest and with our final model after the contest. The model also achieves high level of interpretability evaluated by human experts.

The main contributions of this work are summarized as follows:

1. To the best of our knowledge, AttentionMeSH is the first end-to-end deep learning model with soft-attention mechanism to index MeSH terms in such a large scale (millions of training data). With this relatively simple model, we achieved close to state-of-the-art performance without any sophisticated feature engineering or preprocessing.

2. We develop a novel masking mechanism, which is aimed to handle multi-class classification problems with a large number of classes, like indexing MeSH. We also conduct extensive experiments on how the masking layer settings influence classification performance.

3. We believe AttentionMeSH is the first MeSH annotation model that is capable of providing textual evidence and interpretations of its predictions. We argue that interpretability matters because humans are needed to complete the annotation task.

2 Related Work

2.1 Automatic MeSH Indexing

NLM developed Medical Text Indexer (MTI), a software for providing human indexers with automatic MeSH recommendations (Aronson et al., 2004). MTI takes as input a title and corresponding abstract to generate a set of recommended MeSH terms. MTI has two steps: the first is to generate MeSH candidates for recommendation, and the second is to filter and rank the candidates to give a final output. MTI uses MetaMap and nearest neighbor methods. MetaMap is another NLM-developed tool, which maps mentions in biomedical texts to Unified Medical Language System concepts (Aronson, 2001).

BioASQ is an European Union-funded project that organizes tasks on biomedical semantic indexing and question answering (Tsatsaronis et al., 2015). In the task A of BioASQ, participants are asked to annotate un-indexed PubMed articles with MeSH terms using their models, before they are annotated by the human indexers. The manual annotations are taken as ground truth to evaluate the participating models. DeepMeSH (Peng et al., 2016) is the winner of the latest challenge, BioASQ task 5a, held in 2017. DeepMeSH also uses a two-step strategy: the first step is to generate MeSH candidates and predict the number of output MeSH terms, and the second step is to rank the candidates and take the highest-ranked predicted number of MeSH terms as output. DeepMeSH uses Term Frequency Inverse Document Frequency (TFIDF) and document to vector (D2V) schemes to represent each abstract and generate MeSH candidates using binary classifiers and k-nearest neighbor (KNN) methods over using these features. TFIDF is a traditional weighted bag of word sparse representation of the text and D2V learns a deep semantic representation of the text.

Because state-of-the-art models have less than 0.7 Micro-F, automatic MeSH indexing systems can just serve to assist human indexers. Since human indexers usually add or delete MeSH terms based on the recommendations, *interpretability* of the automatic annotations is very important for them. In this paper we adopt a *local explanation* view of model interpretability (Lipton, 2016), and argue that a good system, in addition to being accurate, should also be able to tell which part of the input supports the indexed MeSH term. This would allow human indexers to be more effective at annotating the article.

2.2 Deep Learning for Text Classification

Automatic indexing of MeSH terms to PubMed articles is a multi-label text classification problem. FastText (Joulin et al., 2016) is a simple and effective method for classifying texts based on n-gram embeddings. (Kim, 2014) used Convolutional Neural Networks (CNNs) for sentence-level classification tasks with state-of-the-art performance on 4 out of 7 tasks they tried. Very deep architectures such as that of (Conneau et al., 2017) have also been proposed for text classification. Motivated by these works we use an RNN-based model for classifying each MeSH term as being a positive label for a given article. We further use attention mechanism to boost performance and provide word-level interpretability.

Recently, there has been work on automatic annotation of International Classification of Diseases codes from clinical texts. (Shi et al., 2017) used character-level and word-level Long Short-Term Memory netowrks to get the document representations and (Mullenbach et al., 2018) used word-level 1-D CNN to get the document representations. Both these works utilized a soft attention strategy where each class gets a specific document represetation by weighted sum of the attention over words or phrases. Mullenbach et al. (2018) also highlighted the need for interpretability when annotating medical texts – in this work we apply similar ideas to the domain of MeSH indexing.

3 Methods

The model architecture is visualized in Figure 1. Starting from an input abstract, title and journal name, words in the document are embedded and fed to BiGRU to derive context-aware representations; KNN-derived articles from training corpus are identified and frequent MeSH terms in them are included as candidate annotations for the document. MeSH terms are embedded, and only those candidates are further considered in attention mechanism. We call it a masking mechanism. We apply an attention mechanism to assign attention weights to each word with respect to each candidate MeSH term, which leads to a MeSH-specific document representation. Finally, we use MeSH-specific document representations as input to perform classifications. For each candidate MeSH term of a document, the model outputs a probability. Binary cross-entropy loss is used for a gradient-based method to optimize the parameters. At inference time, the sigmoid outputs are converted to binary variables by thresholding.

3.1 Document Representation

For each article to be indexed, we first tokenize the journal name, title and abstract to words. In order to use the pre-trained word embeddings[6] provided by BioASQ organizer, we use the same tokenizer as they did. The pre-trained word embeddings are denoted as $\mathbf{E} \in \mathbb{R}^{|\mathcal{V}| \times d_{e1}}$, where $|\mathcal{V}|$ is the vocabulary size and d_{e1} is the embedding size.

We can represent each article by a sequence of word embeddings corresponding to the tokenized text. The word embeddings are initialized by the BioASQ pre-trained word embeddings.

$$\mathbf{D} = \begin{bmatrix} \mathbf{w_1} & ... & \mathbf{w_L} \end{bmatrix}^T \in \mathbb{R}^{L \times d_{e1}},$$

where L is the number of words in the journal name, title and abstract, and $\mathbf{w_i}$ is a vector for word at position i.

For each document representation $\mathbf{D}$, we feed this sequence of word vectors to an BiGRU to derive a context-aware sequence of word vectors:

$$\widetilde{\mathbf{D}} = \text{BiGRU} \left(\mathbf{D} \right) = \begin{bmatrix} \widetilde{\mathbf{w_1}} & ... & \widetilde{\mathbf{w_L}} \end{bmatrix}^T \in \mathbb{R}^{L \times 2d_h},$$

where $\widetilde{\mathbf{w_i}}$ is the corresponding concatenated forward and backward hidden states of each word, and d_h is the hidden size of BiGRU.

---

[6]`http://participants-area.bioasq.org/tools/BioASQword2vec/`

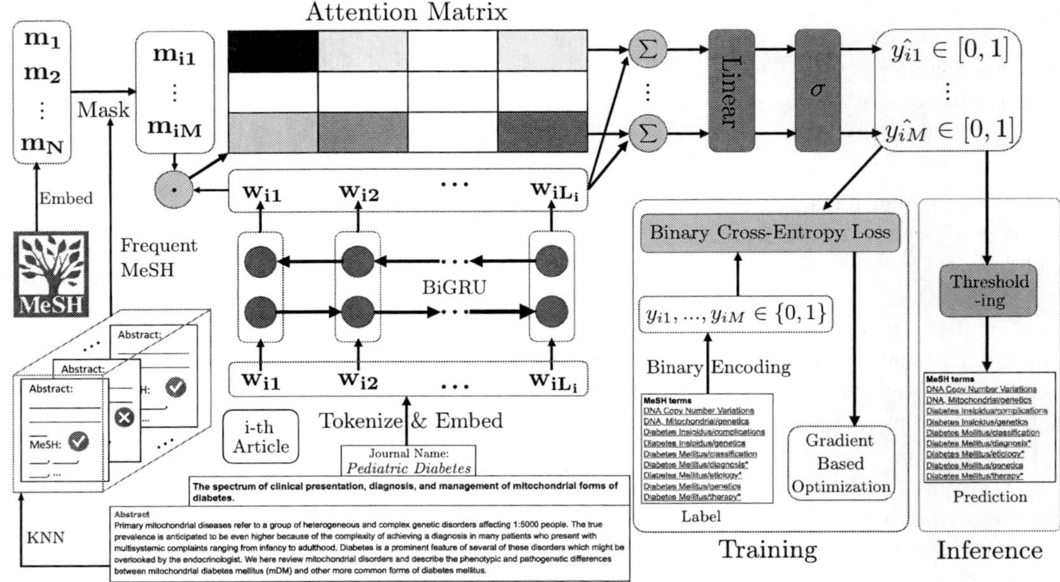

Figure 1: Model Architecture. BiGRU: Bi-directional recurrent gated unit. The example abstract is from (Karaa and Goldstein, 2015).

3.2 MeSH Representation and Masking

We learn the MeSH embedding matrix $\mathbf{H} \in \mathbb{R}^{N \times d_{e2}}$, where N is the number of all MeSH terms (28,340), and d_{e2} is the embedding size. For each article, we consider only a subset of all 28k MeSH terms for two reasons: 1. For each MeSH term, there are far more negative samples than the positive ones. We achieve down-sampling of the negative samples by considering only a subset of all MeSH terms as candidate for each article, so that the classifier only concentrate on choosing a most suitable MeSH among a set of plausible annotations; 2. It's more time efficient than training all the MeSH terms or training the MeSH classifiers one by one. We call it a masking layer.

We use KNN strategy to choose a specific subset of MeSH terms to train for each article:

Each abstract can be represented by IDF-weighted sum of word vectors:

$$\mathbf{d} = \frac{\sum_{i=1}^{n} IDF_i \times \mathbf{w_i}}{\sum_{i=1}^{n} IDF_i} \in \mathbb{R}^{d_e 1},$$

where $\mathbf{w_i}$ is the corresponding word vector, and IDF_i is the inverse document frequency of this word.

We then calculate cosine similarity of representations between the abstracts:

$$\text{Similarity}(i, j) = \frac{\mathbf{d_i^T d_j}}{||\mathbf{d_i}|| \times ||\mathbf{d_j}||}$$

For each article, we find its K nearest neighbors based on cosine similarity. And then we count the MeSH term frequency in these neighbors. The most frequent M MeSH terms are trained for each article. We denote the masked MeSH embedding as $\mathbf{H'}$,

$$\mathbf{H'} = \begin{bmatrix} \mathbf{m_1} & \mathbf{m_2} & ... & \mathbf{m_M} \end{bmatrix} \in \mathbb{R}^{M \times d_{e2}},$$

where we make $d_{e2} = d_h$ so that we could directly get the dot product of each MeSH representation and word vector.

3.3 Attention Mechanism

After getting the document representation and masked MeSH representations, we calculate the dot products between each context-aware word vector and each MeSH embedding, which represents the similarity within each pair:

$$\mathbf{S} = \mathbf{H'} \widetilde{\mathbf{D}}^T = \begin{bmatrix} \widetilde{\mathbf{D}} \mathbf{m_1} & ... & \widetilde{\mathbf{D}} \mathbf{m_M} \end{bmatrix}^T \in \mathbb{R}^{M \times L},$$

We then uses SoftMax function to normalize over the word axis to get attention weights attribution for each MeSH term:

$$\text{SoftMax}(\mathbf{Sim})$$

$$= \begin{bmatrix} \text{SoftMax}(\widetilde{\mathbf{D}}\,\mathbf{m_1}) & ... & \text{SoftMax}(\widetilde{\mathbf{D}}\,\mathbf{m_M}) \end{bmatrix}^T$$

$$= \begin{bmatrix} \boldsymbol{\alpha_1} & ... & \boldsymbol{\alpha_M} \end{bmatrix}^T \in [0,1]^{M \times L},$$

where $\boldsymbol{\alpha}_j \in [0,1]^L$ is the attention weights over words for MeSH term j, and $\sum_{k=1}^{L} \alpha_{jk} = 1$.

3.4 Classification

For each MeSH term, we can have a MeSH-specific representation of document by sum of word vectors weighted by attention weights:

$$\mathbf{R_j} = \boldsymbol{\alpha}_j \widetilde{\mathbf{D}} \in \mathbb{R}^{2d_h},$$

where $\mathbf{R_j}$ is MeSH term j specific document representation. We apply a linear projection layer and sigmoid activatin function to each MeSH term, finally getting the output probability:

$$\hat{y}_j = \sigma(\mathbf{R_j^T}\mathbf{m'_j} + b_j) \in [0,1],$$

where $\mathbf{m'_j}$ and b_j are learnable linear projection parameters for MeSH term j. We model

$$P(\text{MeSH } j \text{ indexed} \,|\, \text{Journal, Title, Abstract}) = \hat{y}_j.$$

3.5 Training

After get the conditioned probability we model, we can calculate the binary cross-entropy loss for each MeSH term:

$$\mathcal{L}_j = -(y_j log(\hat{y}_j) + (1 - y_j)log(1 - \hat{y}_j)),$$

where $y_j \in \{0,1\}$ is the ground-truth label of MeSH j. $y_j = 0$ means MeSH j is not annotated to the article by human indexers, while $y_j = 1$ means MeSH j is annotated. We can get the total loss by summing them up:

$$\mathcal{L} = \frac{1}{M}\sum_{j=1}^{M}\mathcal{L}_j$$

The model is trained end-to-end from word and MeSH embedding to the final projection layer by a gradient-based optimization algorithm to minimize $\mathcal{L}$.

3.6 Inference

At inference time, we will predict the MeSH terms whose predicted probability is larger than a tuned threshold:

$$(\text{predict MeSH } j) = \mathbb{1}(\hat{y}_j > p_j),$$

where p_j is the tuned threshold for MeSH term j. The thresholds are tuned to maximize MiF:

$$p_1, ..., p_N = \underset{p_1,...,p_N}{\operatorname{argmax}} \text{MiF}(\text{Model}, p_1, ..., p_N)$$

We tune p by the the micro-F optimization algorithm described in (Pillai et al., 2013), which they proved to be able to achieve the global maximum.

4 Experiments

4.1 Dataset

We use the dataset provided by BioASQ[7], which contains about 13.5 million manually annotated PubMed articles. The dataset covers 28,340 MeSH terms in total, and each article is annotated 12.69 MeSH terms on average. We selected 3 million articles from 2012 to 2017 for training.

The results reported in this paper are derived from two test sets: **BioASQ Test Sets**: During the challenge, BioASQ provides a test set of several thousands articles each week. **Ours**: we use 100 thousand latest articles to test our model, and all other results are calculated by this dataset. Since our test set is very large, the results will be precise.

4.2 Configuration

The model is implemented using PyTorch (Paszke et al., 2017). The parameter settings are shown in Table 1. We use Adam optimizer and batch size of 32. We train 2 epochs of each model on the 3M article training set, and apply hyperbolic learning rate decay and early stopping strategies (Yao et al., 2007). The training takes 4 days on 2 GPUs (GeForce GTX TITAN X). Before tuning the thresholds for all individual MeSH term, we use a global threshold of 0.35 due to the highly imbalanced dataset.

4.3 Evaluation Metric

The major metric for performance evaluation is Micro-F, which is a harmonic mean of micro-precision (MiP) and micro-recall (MiR) , and is calculated as follows:

$$\text{Micro-F} = \frac{2 \cdot \text{MiP} \cdot \text{MiR}}{\text{MiP} + \text{MiR}},$$

where

$$\text{MiP} = \frac{\sum_{i=1}^{N_a}\sum_{j=1}^{N} y_{ij} \cdot \hat{y_{ij}}}{\sum_{i=1}^{N_a}\sum_{j=1}^{N} \hat{y_{ij}}}$$

[7] http://participants-area.bioasq.org/ general_information/Task6a/

Parameter	Value(s)
$\|\mathcal{V}\|$	1.7M
d_{e1}	256
d_{e2}	512
d_h	256
N	28,340
L	$\leq$512 (truncated if longer)
N_a BioASQ	5,833~10,488
N_a Ours	100,000
K	0.1k, 0.5k, **1k**, 3M
M	128, 256, 512, **1,024**
Learning Rate	0.002, **0.001**, 0.0005
BiGRU Layer(s)	1, 2, **3**, 4

Table 1: Parameter Values. For hyperparameters, we highlight the optimal ones among all tried values.

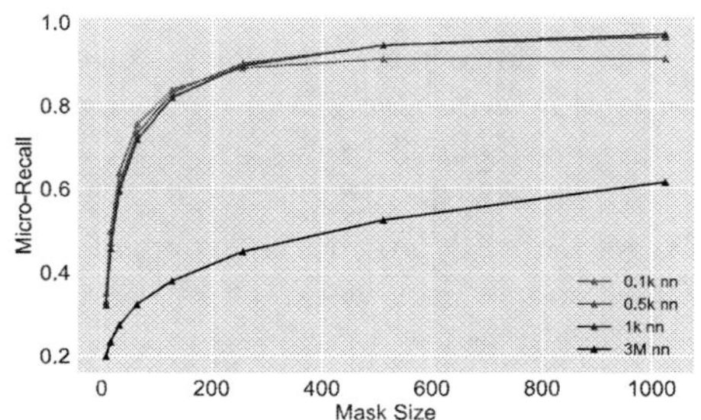

Figure 2: The micro-recall of MeSH terms versus different mask sizes for different numbers of neighbor articles.

$$\text{MiR} = \frac{\sum_{i=1}^{N_a} \sum_{j=1}^{N} y_{ij} \cdot \hat{y_{ij}}}{\sum_{i=1}^{N_a} \sum_{j=1}^{N} y_{ij}}$$

In these equations, i is indexed for articles and j is indexed for MeSH terms, so N_a is the number of articles in the test set, and N is the number of all MeSH terms. y_{ij} and $\hat{y_{ij}}$ are both binary encoded variables to denote whether MeSH term j is in article i in ground-truth and prediction, respectively.

4.4 Evaluation of Masking Layer

Selecting relevant MeSH terms from neighbor articles can be regarded as a weak classifier itself, and high-recall setting is favored in this step. We measure the micro-recall for different masking layer settings, and the results are shown in Figure 2. Basically, there are two hyperparameters for it: the number of neighbor articles K and the number of highest ranking MeSH terms selected M. A non-trivial baseline for K is 3M, i.e. the number of all training articles. Under this circumstance, the ranked MeSH list is determined by global frequency, thus is non-specific to any article.

We choose the number of nearest articles $K = $

Mask Setting	MiP	MiR	MiF
1,024 rd.	0.5891	0.0173	0.0337
1,024 freq.	**0.6863**	0.4257	0.5262
128 n.n.	0.6354	0.5880	0.6108
256 n.n.	0.6690	0.5975	0.6312
512 n.n.	0.6663	0.6116	0.6378
1,024 n.n.	0.6698	**0.6262**	**0.6472**

Table 2: Model Performance with Different Mask Settings. n.n.: MeSH mask selected from nearest neighbor articles ($K = 1000$); freq.: MeSH mask selected from globally frequent MeSH terms; rd.: MeSH mask randomly selected. All results are averaged over models trained by 3 random seeds.

1000 for it gives the highest recalls with the increase of mask size. In fact, micro-recall at $M = 1024$ and $K = 1000$ is about 0.97, which almost guarantees that all true annotations are included as candidate for a document. Before fine-tuning on other hyperparameters and the thresholds of making predictions, we first train the model with different M, and report the results in Table 2.

4.5 Evaluation of Performance

While we were developing the model, we participated in the BioASQ Task6a challenge. During the challenge, there is a test set available each week. Each test set contains several thousands of un-indexed PubMed citations. Each citation has journal name, title, abstract information. Participants will run their models on the test set and upload their predictions of MeSH annotations within a given time. The organizers will then evaluate every participants' predictions and make the results available. The results of the whole Challenges are showed in Figure 3. Furthermore, the results of the last week of the Challenge are showed in Table 3.

Model	Average MiF	Maximum MiF
Access Inn MAIstro	0.2788	0.2788
MeSHmallow	0.3161	0.3161
UMass Amherst T2T	0.4988	0.4988
iria	0.4992	0.5161
MTI First Line Index	0.6332	0.6332
DeepMeSH	0.6451	0.6637
Default MTI	0.6474	0.6474
AttentionMeSH	0.6635	0.6635
xgx	**0.6862**	**0.6880**

Table 3: Model Performance of the Final BioASQ Test Set. The models are ranked top-down from the lowest average MiF to the highest one. Our on-construction AttentionMeSH ranked second by average MiF.

It should be noted that the models we used in

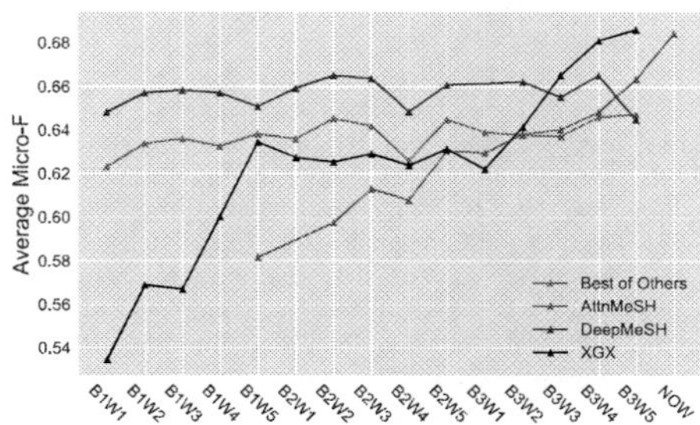

Figure 3: BioASQ Challenge Task6A results. BaWb: Test week b of batch a. From B1W1 to B3W5, we show the average MiF of different models: AttentionMeSH, DeepMeSH, XGX, and the best performance of all other models. Results are retrieved from `http://participants-area.bioasq.org/results/6a/` on June 10th, 2018. And NOW shows the most up-to-date results from our test set.

BioASQ are not our final model. We include the up-to-date results in 'NOW' in Figure 3 and report the ablation test results in Table 4.

Model	MiP	MiR	MiF
AttentionMeSH (AM)	0.6698	0.6262	0.6472
AM w/o BiGRU	0.6362	0.5848	0.6093
AM w/o learning w.e.	0.6657	0.6106	0.6369
AM w/o attention	0.6807	0.5519	0.6095
AM w/ t.t.	0.7048	0.6393	0.6704
AM w/ ensemble & t.t.	**0.7172**	**0.6543**	**0.6844**

Table 4: Model Performance with Ablations and Finer Tuning. w/o: without; w/: with; t.t.: MeSH term threshold tuning; w.e.: word embeddings. Ensembling takes the average prediction of 8 models trained by different seeds. All results, except the ensemble one, are averaged over models trained by 3 random seeds.

4.6 Evaluation of Interpretability

At inference time, attention matrix provides word-level interpretation: For each MeSH prediction, the model shows which words are given high attention. It helps the indexers to evaluate and proof-read the indexing results of our model. Figure 4 shows an example of attention for interpretation.

To qualitatively evaluate the interpretability of different models, the best way would be to measure the time efficiency of manual indexing with the assistance of different models. However, this might require well-trained NLM indexers to evaluate. Instead, we asked two independent researchers with Ph.D. degrees in related fields to label relevant words for 100 MeSH-article pairs. Their intersected labels are regarded as ground-truth. We model the interpretability evaluation as an information retrieval task, and evaluate each method's recall at different numbers of outputs in Table 5. Since other models like DeepMeSH and MTI don't report how to interpret their model outputs, we use string-matching as a non-trivial baseline.

Model	R@5	R@10	R@20
String-Matching	0.3890	0.4180	0.4336
AM Embeddings	0.6180[†]	0.7486[†]	0.8088[†]
AM Whole Model	**0.6929[†‡]**	**0.8389[†‡]**	**0.8993[†‡]**

Table 5: Interpretability Evaluation. R@n: The average recall of ground-truth relevant words if the model outputs n words. AM Whole Model: The whole model of AttentionMeSH is used to get the attention matrix, and n words with highest attention weights will be the output; AM Embeddings: We only use the trained word and MeSH embeddings of AttentionMeSH model, and we output n words that have highest dot products with each specific MeSH. String-Matching: A string matching method that takes all words in the abstracts that are same to any word in the MeSH name. [†]: Significant differences with String-Matching; [‡]: Significant differences with AM Embeddings. Significance is defined by $p < 0.05$ in paired t tests.

5 Discussion

One intrinsic limitation of all present automatic MeSH indexing models, including us, is that these models just annotate MeSH terms from the abstract, title, journal name etc, but they don't look into the article bodies. However, the human indexers in NLM do need to look into the bodies to annotate each article, and thus the textual evidence for certain annotations is missed during training. As such, all present models won't have enough information to do the annotation, and certain percent of false negatives is inevitable, and the performance is upbounded by them. For example, MeSH terms 'Humans', 'Males', 'Females' are annotated to our demo article in Figure 4. However, the abstract doesn't contain any relevant information. 35 articles in our 100 MeSH-article pairs evaluated by experts don't have any words relevant to the MeSH term.

We noted that AttentionMeSH predicted many MeSH terms to documents that were not annotated by NLM indexers, which appears to be "false pos-

PLoS One
Association of SNP rs80659072 in the ZRS with polydactyly in Beijing You chickens

Abstract
The Beijing You chicken is a Chinese native breed with superior meat quality and a unique appearance. The G/T mutation of SNP rs80659072 in the Shh long-range regulator of GGA2 is highly associated with the polydactyly phenotype in some chicken breeds. In the present study, this SNP was genotyped using the TaqMan detection method, and its association with the number of toes was analyzed in a flock of 158 birds of the Beijing You population maintained at the Beijing Academy of Agriculture and Forestry Sciences. Furthermore, the skeletal structure of the digits was dissected and assembled in 113 birds. The findings revealed that the toes of Beijing You chickens were rich and more complex than expected. The plausible mutation rs80659072 in the zone of polarizing activity regulatory sequence (ZRS) in chickens was an essential but not sufficient condition for polydactyly and polyphalangy in Beijing You chickens. Several individuals shared the T allele but showed normal four-digit conformations. However, breeding trials demonstrated that the T allele could serve as a strong genetic marker for five-toe selection in Beijing You chickens.

True Positive MeSH: *Toes*
• …with the number of **toes** was analyzed in a…
• …findings revealed that the **toes** of Beijing You chickens…
• …skeletal structure of the **digits** was dissected and assembled…

True Positive MeSH: *Polymorphism, Single Nucleotide*
• Association of **SNP** rs80659072 in the ZRS…
• …The G/T mutation of **SNP** rs80659072 in the Shh…
• …this **SNP** was genotyped using the TaqMan…

False Positive MeSH: *China*
• …ZRS with polydactyly in **Beijing** You chickens…
• …**Beijing** You chicken is a **Chinese** native breed with…

False Negative MeSH: *Meat*
• …native breed with superior **meat** quality and…
• …The Beijing You **chicken** is a Chinese native breed…
• …in Beijing You **chickens**

Figure 4: Attention Display. In a randomly-selected test article (Chu et al., 2017), we show the 3 words that are given highest attention weights for 4 MeSH terms, including two true positive, one false positive and one false negative predictions.

itives". However, after manual inspection, we noticed that many of our predictions are semantically sensible. For example, both the articles in Figure 4 and Figure 5 discuss genotype-phenotype relationship in Beijing You chickens. However, MeSH term China is annotated to the article in Figure 5, but not the one in Figure 4. We conjecture that this may be due to inconsistency among indexers and that automatic indexing may assign more semantically sensible annotations to enhance the coverage of concepts in a document.

In consideration of the limitations and problems mentioned above, some false positive and false negative MeSH terms are unavoidable. We argue that human experts' performance on test dataset based on the same input as given in BioASQ is needed to provide better evaluation and comparison of performance of current methods.

Concerning how the explanations will help, we just perform a preliminary study by human evaluators, where we model the interpretability as an information retrieval (IR) task. However, the potential users don't regard the annotation task as an IR task. Thus, it would be more convincing to recruit some indexers at NLM and conduct a user study, measuring the annotation efficiency and accuracy with and without the help of AttentionMeSH.

Animal Biotechnology
The effect of a mutation in the 3-UTR region of the HMGCR gene on cholesterol in Beijing-you chickens.

Abstract
The 3-hydroxyl-3-methylglutaryl Coenzyme A reductase (HMGCR) gene was examined for polymorphisms in Beijing-you chickens. A "T" base insert was detected at nucleotide 2749 of the 3-UTR region of the HMGCR gene and was used as the basis for distinguishing a B allele, distinct from the A. Serum and muscle contents of total cholesterol. LDL-cholesterol in serum was significantly lower in AB birds and lowest in BB birds. Real-time PCR showed that the same trends across genotypes occurred in an abundance of HMGCR transcripts in liver, but there was no difference in contents of HMGCR mRNA in breast or thigh muscles. Hepatic expression and serum LDL-cholesterol were meaningfully correlated (partial, with total serum cholesterol held constant, r = 0.923). In muscle, similar genotypic differences were found for the abundance of the LDL receptor (LDLR) transcript. Cholesterol content in breast muscle related to LDLR expression (partial correlation with serum LDL-cholesterol held constant, r = 0.719); the equivalent partial correlation in thigh muscle was not significant. The results indicated that the B allele significantly reduces hepatic abundance of HMGCR transcripts, probably accounting for genotypic differences in serum cholesterol. In muscle, the cholesterol content appeared to reflect differences in LDLR expression with apparent mechanistic differences between breast and thigh.

Figure 5: A Contradictorily Indexed Article (Cui et al., 2010). MeSH term China is annotated to this article, while not to a similar one at Figure 4.

6 Conclusions

We present AttentionMeSH, an automatic MeSH indexer, which is simple and interpretable. It also achieves comparable performance to the current state-of-the-art. Since even the state-of-the-art model has only about 0.69 by MiF metric, manual annotations are still required. Thus, interpretability of the models is vital. We evaluate the interpretability of AttentionMeSH by retrieving capability of experts-labeled relevant words. Our model achieves high performance by this task. To the best of our knowledge, AttentionMeSH is the only interpretable model for indexing MeSH which has close to state-of-the-art performance.

7 Acknowledgement

This work has been supported by NIH grant No. 5R01LM012011. We thank Dr. Chunhui Cai and Dr. Lujia Chen for their independent evaluations of model interpretability. We are also grateful for the anonymous reviewers who gave us very insightful suggestions. The content is solely the responsibility of the authors and does not necessarily represent the official views of the National Institutes of Health.

References

Alan R Aronson. 2001. Effective mapping of biomedical text to the umls metathesaurus: the metamap program. In *Proceedings of the AMIA Symposium*, page 17. American Medical Informatics Association.

Alan R Aronson, James G Mork, Clifford W Gay, Susanne M Humphrey, Willie J Rogers, et al. 2004. The nlm indexing initiative's medical text indexer. *Medinfo*, 89.

Dzmitry Bahdanau, Kyunghyun Cho, and Yoshua Bengio. 2014. Neural machine translation by jointly learning to align and translate. *arXiv preprint arXiv:1409.0473*.

Kyunghyun Cho, Bart Van Merriënboer, Caglar Gulcehre, Dzmitry Bahdanau, Fethi Bougares, Holger Schwenk, and Yoshua Bengio. 2014. Learning phrase representations using rnn encoder-decoder for statistical machine translation. *arXiv preprint arXiv:1406.1078*.

Qin Chu, Zhixun Yan, Jian Zhang, Tahir Usman, Yao Zhang, Hui Liu, Haihong Wang, Ailian Geng, and Huagui Liu. 2017. Association of snp rs80659072 in the zrs with polydactyly in beijing you chickens. *PloS one*, 12(10):e0185953.

Alexis Conneau, Holger Schwenk, Loïc Barrault, and Yann Lecun. 2017. Very deep convolutional networks for text classification. In *Proceedings of the 15th Conference of the European Chapter of the Association for Computational Linguistics: Volume 1, Long Papers*, volume 1, pages 1107–1116.

HX Cui, SY Yang, HY Wang, JP Zhao, RR Jiang, GP Zhao, JL Chen, MQ Zheng, XH Li, and J Wen. 2010. The effect of a mutation in the 3-utr region of the hmgcr gene on cholesterol in beijing-you chickens. *Animal biotechnology*, 21(4):241–251.

Armand Joulin, Edouard Grave, Piotr Bojanowski, and Tomas Mikolov. 2016. Bag of tricks for efficient text classification. *arXiv preprint arXiv:1607.01759*.

Amel Karaa and Amy Goldstein. 2015. The spectrum of clinical presentation, diagnosis, and management of mitochondrial forms of diabetes. *Pediatric diabetes*, 16(1):1–9.

Yoon Kim. 2014. Convolutional neural networks for sentence classification. *arXiv preprint arXiv:1408.5882*.

Yann LeCun, Yoshua Bengio, and Geoffrey Hinton. 2015. Deep learning. *nature*, 521(7553):436.

Zachary C Lipton. 2016. The mythos of model interpretability. *arXiv preprint arXiv:1606.03490*.

James G Mork, Antonio Jimeno-Yepes, and Alan R Aronson. 2013. The nlm medical text indexer system for indexing biomedical literature. In *BioASQ@ CLEF*.

James Mullenbach, Sarah Wiegreffe, Jon Duke, Jimeng Sun, and Jacob Eisenstein. 2018. Explainable prediction of medical codes from clinical text. *arXiv preprint arXiv:1802.05695*.

Adam Paszke, Sam Gross, Soumith Chintala, Gregory Chanan, Edward Yang, Zachary DeVito, Zeming Lin, Alban Desmaison, Luca Antiga, and Adam Lerer. 2017. Automatic differentiation in pytorch.

Shengwen Peng, Ronghui You, Hongning Wang, Chengxiang Zhai, Hiroshi Mamitsuka, and Shanfeng Zhu. 2016. Deepmesh: deep semantic representation for improving large-scale mesh indexing. *Bioinformatics*, 32(12):i70–i79.

Ignazio Pillai, Giorgio Fumera, and Fabio Roli. 2013. Threshold optimisation for multi-label classifiers. *Pattern Recognition*, 46(7):2055–2065.

Mike Schuster and Kuldip K Paliwal. 1997. Bidirectional recurrent neural networks. *IEEE Transactions on Signal Processing*, 45(11):2673–2681.

Haoran Shi, Pengtao Xie, Zhiting Hu, Ming Zhang, and Eric P Xing. 2017. Towards automated icd coding using deep learning. *arXiv preprint arXiv:1711.04075*.

George Tsatsaronis, Georgios Balikas, Prodromos
Malakasiotis, Ioannis Partalas, Matthias Zschunke,
Michael R Alvers, Dirk Weissenborn, Anastasia
Krithara, Sergios Petridis, Dimitris Polychronopoulos, et al. 2015. An overview of the bioasq
large-scale biomedical semantic indexing and question answering competition. *BMC bioinformatics*,
16(1):138.

Yuan Yao, Lorenzo Rosasco, and Andrea Caponnetto.
2007. On early stopping in gradient descent learning. *Constructive Approximation*, 26(2):289–315.

Extraction Meets Abstraction:
Ideal Answer Generation for Biomedical Questions

Yutong Li,[*] Nicholas Gekakis,[*] Qiuze Wu,[*] Boyue Li,[*]
Khyathi Raghavi Chandu, Eric Nyberg
Language Technologies Institute, Carnegie Mellon University
{anareshk,hkesavam,madhurad,pkalwad,kchandu,teruko,ehn}@cs.cmu.edu

Abstract

The growing number of biomedical publications is a challenge for human researchers, who invest considerable effort to search for relevant documents and pinpointed answers. Biomedical Question Answering can automatically generate answers for a user's topic or question, significantly reducing the effort required to locate the most relevant information in a large document corpus. Extractive summarization techniques, which concatenate the most relevant text units drawn from multiple documents, perform well on automatic evaluation metrics like ROUGE, but score poorly on human readability, due to the presence of redundant text and grammatical errors in the answer. This work moves toward abstractive summarization, which attempts to distill and present the meaning of the original text in a more coherent way. We incorporate a sentence fusion approach, based on Integer Linear Programming, along with three novel approaches for sentence ordering, in an attempt to improve the human readability of ideal answers. Using an open framework for configuration space exploration (BOOM), we tested over 2000 unique system configurations in order to identify the best-performing combinations for the sixth edition of Phase B of the BioASQ challenge.

1 Introduction

Human researchers invest considerable effort when searching very large text corpora for answers to their questions. Existing search engines like PubMed (Falagas et al., 2008) only partially address this need, since they return relevant documents but do not provide a direct answer for the user's question. The process of filtering and combine information from relevant documents to obtain an ideal answer is still time consuming (Tsatsaronis et al., 2015). Biomedical Question Answering (BQA) systems can automatically generate ideal answers for a user's question, signif-

icantly reducing the effort required to locate the most relevant information in a large corpus.

Our goal is to build an effective BQA system to generate coherent, query-oriented, non-redundant, human-readable summaries for biomedical questions. Our approach is based on an extractive BQA system (Chandu et al., 2017) which performed well on automatic metrics (ROUGE) in the 5th edition of the BioASQ challenge. However, owing to the extractive nature of this system, it suffers from problems in human readability and coherence. In particular, extractive summaries which concatenate the most relevant text units from multiple documents are often incoherent to the reader, especially when the answer sentences jump back and forth between topics. Although the existing extractive approach explicitly attempts to reduce redundancy at the sentence level (via SoftMMR), stitching together existing sentences always admits the possibility of redundant text at the phrase level. We improve upon the baseline extractive system in 3 ways: (1) re-ordering the sentences that are selected by the extractive algorithm; (2) fusing words and sentences to form a more humanireadable summary; and (3) using automatic methods to explore a much larger space of system configurations and hyperparameter values when optimizing system performance. We hypothesize that the first two techniques will improve the coherence and human readability, while the third technique provides an efficient framework for tuning these approaches in order to maximize automatic evaluation (ROUGE) scores.

2 Overview of Baseline System Architecture

In this section, we provide a brief layout of our baseline system, which achieved the top ROUGE scores in the final test batches of the fifth edition of BioASQ Challenge (Chandu et al., 2017). This system includes baseline modules for relevance ranking, sentence selection, and sentence tiling.

The baseline relevance ranker performs the fol-

---

[*] denotes equal contribution

Proceedings of the 2018 EMNLP Workshop BioASQ: Large-scale Biomedical Semantic Indexing and Question Answering, pages 57–65
Brussels, Belgium, November 1st, 2018. ©2018 Association for Computational Linguistics

lowing steps: 1) Expand concepts in the original question using a metathesaurus, such as UMLS (Bodenreider, 2004) or SNOMEDCT (Donnelly, 2006); and 2) calculate a relevance score (e.g. Jaccard similarity) for each question/snippet pair (to measure relevance) and each pair of generated snippets (to measure redundancy). The baseline sentence selection model used the Maximal Marginal Relevance (MMR) algorithm (Carbonell and Goldstein, 1998), which iteratively selects answer sentences according to their relevance to the question and their similarity to sentences that have already been selected, until a certain number of sentences have been selected. The baseline sentence tiling module simply concatenates selected sentences up to a given limit on text length (200 words), with no attempt to module or improve the coherence of the resulting summary.

The baseline system achieved high ROUGE scores, but performed poorly on the human readability evaluation in BioASQ 2017. In order to improve human readability, we first developed several post-processing modules, such as sentence reordering and sentence fusion, which will be discussed in detail in following sections.

3   Sentence Ordering

3.1   Motivation

As discussed in Section 1, we tried to improve upon the Soft MMR system (Chandu et al., 2017). This pipeline assumes the relevance to be a proxy for ordering the selected sentences to generate the final summary. On the other hand, it does not take into account the flow and transition of sentences to build a coherent flow between these sentences. Since the maximum length of the answer is 200 words (as imposed by the guidelines of the competition), this system optimizes on selecting the most non-redundant query relevant sentences to maximize the ROUGE score. In this section, we focus on different types of sentence ordering that lead to more coherent answers.

3.2   Algorithms and Techniques

3.2.1   Similarity Ordering

The intuition behind the Similarity Ordering algorithm is that sentences that have similar content should appear consecutively so that the generated answer is not jumping back and forth between topics. Our implementation is based on work by Zhang (2011), which discusses the use of similarity metrics at two levels - first to cluster sentences, and then to order them within a cluster - which can lead to big improvements in coherency and readability. We apply this approach to the BQA

domain, where we cluster our set of candidate answers using k-means with $k = 2$. We then order the sentences within each cluster, starting with the candidate sentence nearest to the centroid of its cluster and working outward. The intuition is that the most central sentence will contain the largest number of tokens shared by all the sentences in the cluster, and is therefore likely to be the most general or comprehensive sentence in the cluster. This supports our goal of an ideal answer that begins with a broad answer to the question, followed by specifics and supporting evidence from the literature.

In Figure 1a we see that the order of the sentences that appear in the final answer is completely independent of their ordering in the original snippets.

3.2.2   Majority Ordering

The Majority Ordering algorithm Barzilay and Elhadad (2002) makes two main assumptions that are quite reasonable: sentences coming from the same parent document should be grouped together, and the most coherent ordering of a group of sentences is how they were presented in their parent document. Topically, it is logical that sentences drawn from the same parent document would be similar. Grammatically and syntactically, it is logical that the sentences may be structured in a way such that maintaining an invariant ordering would augment human comprehension.

Specifically, the Majority Ordering algorithm groups sentences by their parent document and then orders the blocks by the ranking of the highest ranked sentence in a block. Figure 1 illustrates the differences between Similarity Ordering, Majority Ordering, and Block Ordering. The color of each sentence unit indicates the document it was selected from, and the suffix indicates the relevance score of that unit within the document.

3.2.3   Block Ordering

Intuitively, the Block Ordering algorithm is an amalgamation of the Similarity Ordering and Majority Ordering algorithms. The Block Ordering algorithm has two primary components. The first component involves grouping the sentences into blocks based on their parent document. This step is shared between the Block Ordering algorithm and the Majority Ordering algorithm. The second step involves ordering the grouped blocks of text.

The algorithm for ordering the blocks of texts combines document heuristics with our Similarity Ordering algorithm. We first order the blocks by their length (the number of sentences in teh block). For blocks of equal length, we calculate the similarity of each block with the last fixed sentence. Hence, given the last sentence of the preceding block, we select the next block first by its length, and then by the similarity of the block with the preceding sentence. If there is no single longest block to begin the answer, then we select the longest block that is most similar to the entire answer. This algorithm is tuned for specific goals with respect to human comprehension and readability. Grouping the sentences into blocks is done to maximize local coherence. The use of block length as an ordering heuristic is done to order topics by relevance. Finally, ordering blocks of equal length by similarity to the preceding sentence is done to maximize sentence continuity and fluidity.

In Figure 1c the green block is ordered first because it is the longest. The blue block is ordered second because it has the highest similarity score with sentence 3.4. The yellow block is ordered third because it has a higher similarity with sentence 2.2, and the red block is thus last.

3.3 Quantitative Analysis

To evaluate our approaches, we performed a manual analysis of 100 different answers, ordered by each of our proposed ordering algorithms (see Table 1). We rate each ordering as 'reasonable' or 'unreasonable'. Note that this rating does not pass judgment on the correctness of the answer, since it is designed for a comparative analysis at the module level (i.e. to compare ordering approaches rather than content selection).

Algorithm	Reasonable	Unreasonable
Baseline	59	41
Similarity Ordering	55	45
Majority Ordering	71	29
Block Ordering	75	25

Table 1: Manual evaluation of sentence ordering

3.4 Qualitative Analysis

Because sentence ordering in the baseline system is based solely on question-answer relevance, we identified two major issues: global coherence and local coherence.

The global coherence issue is generally a problem of layout and cohesiveness. An ideal answer would begin with a broad answer to the question and move into more specific details and any available evidence to support the answer. Further, an ideal answer should not be hopping back and forth between topics and should stick to one before moving on to another. The baseline system did a decent job of beginning with a broad answer to the question because the input sequence is ordered by their relevance score. However after the

first sentence, answers tended towards redundant information and divergent trains of thought.

The local coherence issue has more to do with the semantics of the sentence and grammatical restrictions of the language. For instance, language like 'There was also' should not appear as the first sentence in an answer because this makes no sense logically. Additionally certain words like 'Furthermore' indicate that the content of the sentence is highly dependent on the content of the preceding sentence(s), and this dependency is frequently broken by the baseline ordering approach.

3.4.1 Similarity Ordering

We found that the Similarity Ordering performed poorly; only 55 of 100 answers were deemed 'reasonable'. We believe that this is due to the high degree of similarity between the candidate sentences in our domain. Because the candidate sentences are so similar to each other, the results of clustering are highly variant and appeared to be almost arbitrary at times. All the sentences contain similar language and key phrases that makes it difficult to create meaningful sub-clusters. Additionally, one of the biggest problems with our system is due to the sentences that began with phrases like 'However' and 'Furthermore' that place strict requirements on the content of the preceding sentence. This was particularly problematic for the Similarity Ordering algorithm which has no mechanism for making sure that such sentences are placed logically with their dependent sentences. The Similarity Ordering algorithm does perform relatively well in creating logical groups of sentences that cut down on how often an answer is jumping from one topic to another. Additionally these groups are ordered well, beginning with the more general of the two and then finishing with specifics and a presentation of the supporting data. However, we note that the problems with local coherence greatly outweigh the strengths in global coherence since a good answer can still be coherent, even if the organization could be improved, whereas if local coherence is poor, then the answer becomes nonsensical.

3.4.2 Majority Ordering

The Majority Ordering algorithm proved to be a successful method for ordering sentences, where 71 out of 100 answers were deemed 'reasonable'. The Majority Ordering displayed very strong local coherence, which confirms the hypothesis that sentences should likely be kept in their original ordering to maximize human readability and coherence.

However, this algorithm faced issues with global coherence. It produced answers that start

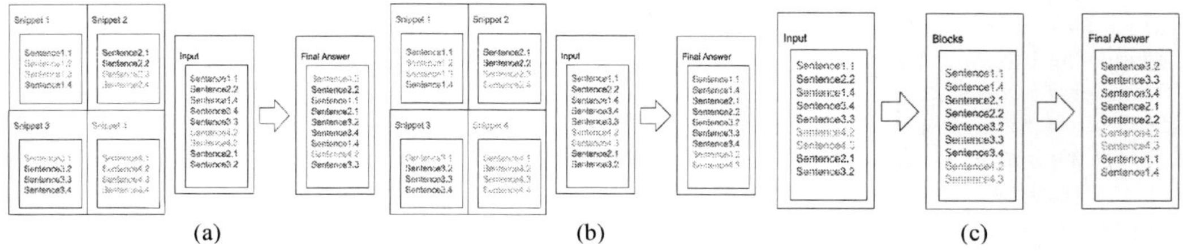

(a)   (b)   (c)

Figure 1: (a) Similarity Ordering (b) Majority Ordering (c) Block Ordering

with a relevant topic more often than not; however, after the initial block, it struggled to smoothly transition from one block to the next. This is consistent with expectations for the Majority Ordering algorithm. The block with highest rated sentence is ordered first, which explains why the first block is frequently the most topically relevant. After the initial block placement, however, the algorithm makes no explicit attempts to manage or smooth transitions between blocks. Compared with the other two algorithms, this is where the Majority Ordering algorithm displays its poorest performance. It performs strongly when ordering sentences within a block, enforcing local coherence so that sentences beginning with language such as 'Finally', 'Lastly', 'Therefore', etc. followed a related sentence that satisfied the sequential dependency.

3.4.3 Block Ordering

The Block Ordering algorithm produced the best answers, with 75 out of 100 answers ranked as 'reasonable'. This is consistent with our expectations, as the Block Ordering algorithm effectively combines the strongest aspects of the Majority Ordering and Similarity Ordering algorithms. With respect to local coherence, this algorithm displays similar performance when compared to the Majority Ordering algorithm, while displaying stronger coherence between blocks (due to the use of a similarity metric to order blocks). This algorithm also displayed the strongest global coherence, which is likely due to first grouping the sentences into blocks and then ordering them.

This algorithm displayed one core weakness, which is its inability to identify high-quality opening sentences. This is due to the usage of block length as a heuristic for topic relevance. While in the majority of cases this heuristic proved to be successful, accounting for these outliers may significantly improve the performance of the Block Ordering algorithm. We note that the Block ordering algorithm performed well in producing high-quality, coherent answers; although the develop-

ment of coherence models and measures is not the main focus of this paper, we can see that Block Ordering performs the best with respect to the simple coherence evaluation we conducted.

4 Sentence Fusion

An observed weakness of the original system is that the generated summaries often contain highly repetitive information. While MMR is added in the pipeline to deal with redundancy and maximize the diversity of covered information, extractive summarization still picks entire sentences that may partially overlap with a previously selected sentence. To tackle this problem, we introduce sentence fusion as a way to identify common information among sentences and apply simple abstractive techniques over the baseline extractive summaries.

4.1 Methodology

Given a set of candidate sentences generated by the pipeline for each summary, the sentence fusion module operates in two steps: 1) the candidate set is expanded to include fused sentences, and 2) sentences are selected from the expanded set to produce a new summary.

4.1.1 Expansion of Candidate Set

To generate fused sentences, we begin by building upon previous work on multiple-sentence compression (Filippova, 2010), in which a directed word graph is used to express sentence structures. The word graph is constructed by iteratively adding candidate sentences. All words in the first sentence are added to the graph by creating a sequence of word nodes. A word in the following sentence is then mapped onto an existing word node if and only if it is the same word, with the same part of speech. Our assumption is that a shared node in the word graph is likely to refer to the same entity or event across sentences.

We then find a K-possible fused sentence by searching for the K-shortest path within the word

60

graph. Definition of the edge weights follows from the original paper (Filippova, 2010):

$$w(e_{ij}) = \frac{\frac{freq(i)+freq(j)}{\sum_{s \in S} diff(s,i,j)^{-1}}}{freq(i) \times freq(j)}$$

where $diff(s, i, j)$ is the difference between the offset positions of word i and j in sentence s. Intuitively, we want to promote a connection between two word nodes with close distance, and between nodes that have multiple paths between them. We also prefer a compression path that goes through the most frequent no-stop nodes to emphasize important words.

When applying the sentence fusion technique to the BioASQ task, we first pre-process the candidate sentences to remove transition words like 'Therefore' and 'Finally'. Such transition words may be problematic because they are not necessarily suitable for the new logical intent in fused sentences, and may break the coherence of the final answer. We also constrain fusion so that the fused sentences are more readable. For instance, we only allow fusing of pairs of sentences that are of proper length, in order to avoid generating overly complicated sentences. We also avoid fusing sentences that are too similar or too dissimilar. In the first case, information in the two sentences is largely repetitive, so we simply discard the one containing less information. In the latter case, fusing two dissimilar sentences more likely confuses the reader with too much information rather than improving the sentence readability. Finally, we add a filter to discard ill-formed sentences, according to some hand-crafted heuristics.

4.1.2 Selecting Sentences from Candidate Set

The next step is to select sentences from the candidate set and produce a new summary. An Integer Linear Program (ILP) problem is formulated as follows, according to (Gillick and Favre, 2009):

$$\max_{y,z} \sum_{i=1}^{N} w_i z_i, \text{ such that } \sum_{j=1}^{M} A_{ij} y_j \geq z_i, A_{ij} y_j \leq z_i,$$

$$\sum_{j=1}^{M} l_j y_j \leq L, y_j \in \{0,1\}, z_i \in \{0,1\}$$

In the equation, z_i is an indicator of whether concept i is selected into the final summary, and w_i is the corresponding weight for the concept. The goal is to maximize the coverage of important concepts in a summary. During the actual experiments, we assign diminishing weights so that later occurrences of an existing concept are less important. This forces the system to select a more diverse set of concepts. We follow the convention of using bigrams as a surrogate for concepts (Taylor Berg-Kirkpatrick and Klein, 2011; Dan Gillick and Hakkani-Tur, 2008), and bigram counts as initial weights. Variable A_{ij} indicates whether concept i appears in sentence j, and variable y_j indicates if a sentence j is selected or not.

4.2 Discussion

Table 2 shows the results of different configurations of the ordering and fusion algorithms (Rows 1 - 4, Row 7, Row 9). Though the overall ROUGE score drops slightly from 0.69 to 0.61 after sentence fusion with the ILP-selection step, this is still competitive with other systems (including the baseline). The sentence re-ordering does not directly impact the ROUGE scores.

We manually examined the fused sentences for 50 questions. We found that our sentence fusion technique is capable of breaking down long sentences into independent pieces, and is therefore able to disregard irrelevant information. For example, given a summary containing the original sentence:

'Thus, miR-155 contributes to Th17 cell function by suppressing the inhibitory effects of Jarid2. (2014) bring microRNAs and chromatin together by showing how activation-induced miR-155 targets the chromatin protein Jarid2 to regulate proinflammatory cytokine production in T helper 17 cells.'

our fusion technique is able to extract important information and formulate it into complete sentences, producing a new summary containing the following sentence:

'Mir-155 targets the chromatin protein jarid2 to regulate proinflammatory cytokine expression in th17 cells.'

The fusion module is also able to compress multiple sentences into one, with minor grammatical errors. For example:

Sentence 1: *'The RESID Database is a comprehensive collection of annotations and structures for protein post-translational modifications including N-terminal, C-terminal and peptide chain cross-link modifications[1].'*

Sentence 2: *'The RESID Database contains supplemental information on post-translational modifications for the standardized annotations appearing in the PIR-International Protein Sequence Database[2]'*

our approach produces the fused sentence:

'The RESID Database contains supplemental information on post-translational modifications[1] is a comprehensive collection of annotations and structures for protein post-translational modifications including N-terminal, C-terminal and peptide chain cross-link modifications[2].'

However, the overall quality of fused sentences is not stable. As shown in Figure 2, around 25% of the selected sentences in final summaries are

	Method	Rouge-2	Rouge-SU4	Avg Precision	Avg Recall	Avg F1	Avg Length
1	**Baseline System**	0.6948	0.6890	0.2297	0.8688	0.3207	173.31
2	**MMR + Order**	0.6291	0.6197	0.2758	0.8118	0.3633	140.39
3	**MMR + Fusion**	0.6183	0.6169	0.2783	0.8094	0.3687	139.24
4	**MMR + Relevance + Order**	0.6357	0.6256	0.2728	0.8124	0.3606	143.55
5	**MMR + Relevance + Order + Post**	**0.6215**	**0.6126**	**0.2788**	**0.8111**	**0.3668**	**139.10**
6	**MMR + Relevance + Order + Fusion + LM**	0.6114	0.6042	0.2775	0.8113	0.3682	141.21
7	**MMR + Relevance + Order + Fusion**	0.6213	0.6101	0.2686	0.8099	0.3579	143.94
8	**MMR + Relevance + Order + Fusion + Post**	0.6017	0.5932	0.2775	0.8091	0.3653	140.38
9	**MMR + Fusion + Order**	0.6223	0.6159	0.2840	0.8181	0.3745	138.79
10	**MMR + Fusion + Relevance + Order**	**0.6257**	**0.6214**	**0.2825**	**0.8193**	**0.3730**	**139.73**
11	**MMR + Fusion + Relevance + Order + Post**	0.6149	0.6096	0.2886	0.8126	0.3768	136.43
12	**Fusion + MMR + Relevance + Order**	0.6112	0.6103	0.2837	0.8211	0.3723	142.11
13	**Fusion + MMR + Relevance + Order + Post**	0.6048	0.6040	0.2898	0.8143	0.3789	137.78

Table 2: Performance of different module combinations on Test Batch 4, BioASQ 4th edition.

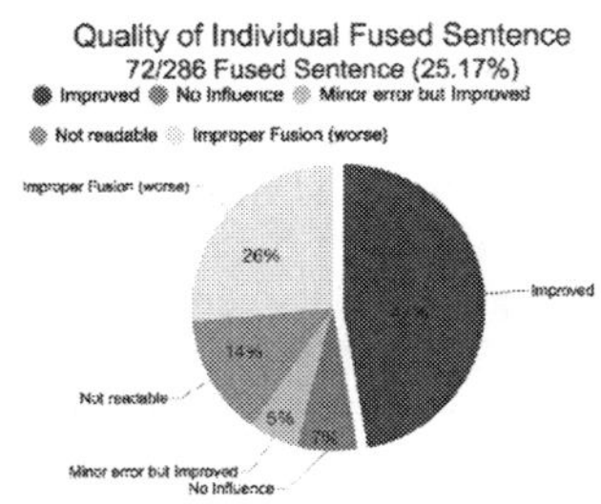

Figure 2: Quality of Fused Sentences

fused. Among the fused sentences, 47% improved the overall readability by reducing redundancy and repetition. 5% of the sentences have improved readability with minor grammatical errors, such as a missing subordinate conjunction or superfluous discourse markers. 8% of the fused sentences did have an appreciable effect on readability. However, a large number of fused sentences (around 26 %) were not coherent and degraded the quality of the answer.

5   Further Improvements

In order to further improve the performance of our system, we made a few modifications to each module in the system, and improved the overall architecture of the module pipeline:

· **Modification of System Architecture:** We intuited that the ILP process in the sentence fusion model could not handle a very large number of candidate inputs, producing a lot of (redundant, similar) fused sentences. In order to resolve this problem, we removed the ILP model from the sentence fusion step, and moved the sentence fusion step before the sentence selection module (Rows 12-13), so that the MMR algorithm in the sentence selection module could take care of eliminating redundant fused sentences.

· **Modifications to Sentence Selection Module and Relevance Ranker:** For the sentence selection module, we modified the original MMR model. The original MMR model selected a fixed number of sentences, which naturally introduced repetition. In order to reduce repetition, we built a so called '*Early-Stop MMR*' which stops selecting sentences when maximum overlap score grows beyond a certain threshold and minimum relevance score drops down below another threshold (Rows 4-8).

For the relevance ranker, we explore an alternative similarity metric ((Row 6). The Query Likelihood Language Model (Schütze et al., 2008) is widely used in information retrieval. We formulated the relevance ranking procedure as an information retrieval problem and used a language model, so that long sentences would get higher penalty.

· **Post-Processing:** To further reduce repetition, we add an additional filter before final concatenation by iteratively adding the selected sentences to the final output, and discarding a sentence if it is too similar to the existing summary (Rows 8,11 and 13se).

6   Configuration Space Exploration

Configuration Space Exploration (CSE) is the technique of trying different combinations of configurations of all modules to find the best configuration (Yang et al., 2013; Yang, 2016). We used the BOOM framework to explore and optimize the space of hyperparameters and module configurations. We explored 2,268 unique configurations of three different hyperparameters: α, used for the MMR module; k, used for the clustering-based Ordering module; and a token limit, used in the Tiling module. Figure 3 shows the pipeline structure we used.

· **Alpha:** This parameter of the MMR module

	Question	Which syndrome is associated with mutant DVL1?
	Ideal Answer	**Mutations in DVL1 cause an osteosclerotic form of Robinow syndrome.**
1	**MMR + Relevance + Order + Fusion**	We identified de novo frameshift mutations in DVL1, a mediator of both canonical and non-canonical Wnt signaling, as the cause of RS-OS, an RS subtype involving osteosclerosis, in three unrelated individuals. Argeted Sanger sequencing in additional subjects with DRS uncovered DVL1 exon 14 mutations in five individuals, including a pair of monozygotic twins. DVL1 frameshift mutations clustering in the penultimate exon cause autosomal-dominant Robinow syndrome. **Mutations in DVL1 cause an osteosclerotic form of Robinow syndrome.**
2	**MMR + Relevance + Fusion + Order**	**Mutations in DVL1 cause an osteosclerotic form of Robinow syndrome.** DVL1 frameshift mutations clustering in the penultimate exon cause autosomal-dominant Robinow syndrome. We identified de novo frameshift mutations in DVL1, a mediator of both canonical and non-canonical Wnt signaling, as the cause of RS-OS, an RS subtype involving osteosclerosis, in three unrelated individuals. Argeted Sanger sequencing in additional subjects with DRS uncovered DVL1 exon 14 mutations in five individuals, including a pair of monozygotic twins.
3	**Fusion + MMR + Relevance + Order + Post**	**DVL1 frameshift mutations in DVL1 cause an osteosclerotic form of Robinow syndrome.** We identified de novo frameshift mutations in DVL1, a mediator of both canonical and non-canonical Wnt signaling, as the cause of RS-OS, an RS subtype involving osteosclerosis, in three unrelated individuals. Argeted Sanger sequencing in additional subjects with DRS uncovered DVL1 exon 14 mutations in five individuals, including a pair of monozygotic twins.

Table 3: System performance comparing Fusion + Ordering and Ordering + Fusion

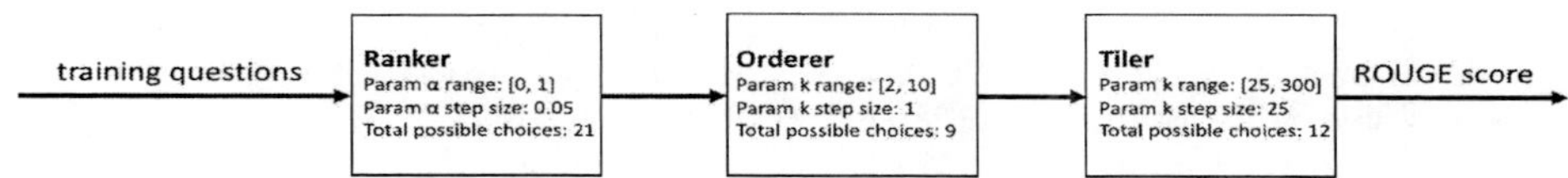

Figure 3: Structure of the CSE pipeline.

controls the trade-off between snippet similarity to the question and snippet similarity to already selected snippets. In our experiments alpha was varied between 0 and 1 at intervals of .05. We have found that the ideal value for alpha is 0.1.

- **Number of Clusters:** The k in the Ordering module controls the number of clusters used to order the snippets for the clustering-based Sentence Ordering algorithms. A small k value produces few, general clusters, while a large k value produces many highly specific clusters with the danger of creating clusters that are actually meaningless or having many clusters that contain a single sentence. In our experiments, k was tested at values from 2 to 10. Although the effect on Rouge score was very small, we have found that the ideal value for k is 3. A caveat to this result is that we are measuring the effect hyperparameter k has on the final Rouge scores achieved by the system. Since the purpose of k is to assist in sentence ordering, not precision or recall, we would expect that adjusting k would have a negligible impact on the Rouge score. Further parameter tuning is needed in cases like this where the primary effect of the parameter is not easily captured by Rouge.

- **Token Limit:** The token limit is used by the Tiling module to set a maximum number of allowed tokens in the answer. If the cumulative token count of the selected snippets exceeds the token limit then sentences will be removed from the end of the final answer until the token limit is satisfied. In our experiments the token limit was tested at values from 25 to 300 in increments of 25. We have found that the ideal value for the token limit is 100.

The two distinct clusters found in the histogram shown in Figure 4 are entirely explained by the token limit. All scores less than 0.27 were obtained by configurations where the token limit was set to 25. The rest of the scores, all above 0.28 were obtained by configurations where the token limit was greater than or equal to 50. In addition to the Rouge score penalty for extremely low token limits, we observed a significant, though much smaller, penalty for token limits of 150 and greater.

7 Results

7.1 Analysis: Effects of Individual Modules

Table 2 shows the results of extensions to the baseline system. Two systems are highlighted (Rows 5 and 10)), as they give the most balanced results between the quality of retrieved information and conciseness: one system performs sentence selection, then ranks sentences prior to ordering by relevance, and applies the additional post-processing step (Row 5); the other system performs sentence

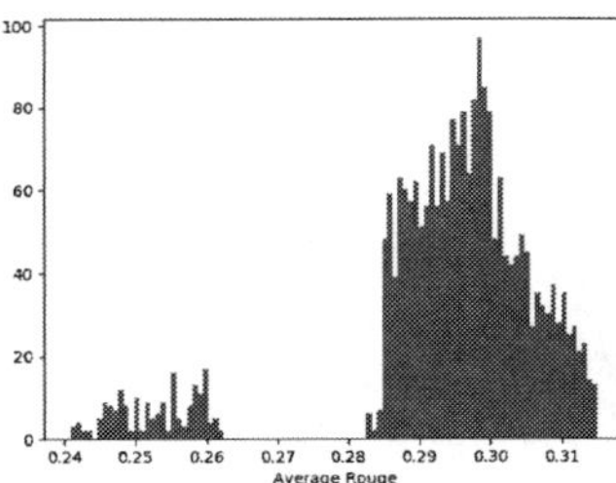

Figure 4: Distribution of Rouge scores across all hyperparameter configurations.

selection, fusion, and then ranks sentences prior to ordering without the post-processing step (Row 10).

Rows 5, 8, 11 and 13 show the effectiveness of the additional post-processing step. Overall, this procedure is able to reduce the answer length, while preserving important information. We observed that the post-processing step is less effective when fusion is performed after MMR. This is because in these settings, there is an additional sentence selection step in the fusion module using integer linear programming that forces the selected sentences to be diverse. In all other settings, including when fusion is performed prior to MMR, we only have one sentence selection step. Since MMR iteratively selects sentences according to both similarity and relevance, the last selected ones may be informative but repetitive. Row 6 shows our experiments with language modeling; the language model gives a higher penalty to longer sentences, which produces shorter but less informative results.

7.2 Analysis: Impact of System Architecture

Exploring the performance of systems using different architectures, we observed that systems with fusion prior to ordering can generate more logically coherent summaries. Table 3 shows an example. All underlined sentences express the same fact that DVL1 is the cause of Robinow syndrome. In Row 1, where fusion is performed after ordering, there is a sentence that serves like an explanation between the underlined sentences, which breaks the logical coherence. In Row 2 and Row 3 where ordering is performed after fusion, the generated answers demonstrate better coherence: All underlined sentences are placed together, following by the explanation; The opening sentences are also more concise and more directly related to the question.

We also experimented with architectures where the fusion module is run prior to MMR, and MMR is used as the only sentence selection step. In these systems, MMR receives many fused sentences that overlap and complement each other at the same time, because all similar sentences are fused prior to sentence selection. As a result, such architectures sometimes produce summaries that are more repetitive compared to others.

8 Conclusion and Future Work

Though extractive summarization techniques can be developed to maximize performance as measured by evaluation metrics like ROUGE, such systems suffer from human readability issues as mentioned above. In this paper we attempted to combine extractive techniques with simple abstractive extensions, by extracting the most relevant non-redundant sentences, re-ordering and fusing them to make the resulting text more human-readable and coherent. Using an initial set of 100 candidate answer sets, we experimented with different ordering algorithms such as Similarity, Majority and Block Ordering, and identified that Block Ordering performs better the others in terms of global and local coherence. We then introduced an Integer Linear Programming based fusion module that is capable of not only fusing repeated content, but also breaks down complicated sentences into simpler sentences, thus improving human readability. The improved baseline system achieved a ROUGE-2 of 0.6257 and ROUGE-SU4 of 0.6214 on test batch 4 of BioASQ 4b. We acknowledge that providing immediate human feedback during the BioASQ competition is manually expensive, although this would greatly help in tuning our systems. We were able to perform a manual evaluation on a sub-sample of the data, in order to introduce the use of human evaluation during system development. We also incorporated an automatic evaluation framewook (BOOM) which allowed us to test many different system configurations and hyperparameter values during system development. As BOOM is completely general and can be applied to any pipeline of Python modules, this adaptation was relatively straightforward, and allowed us to automatically test more than 2,000 different system configurations.

In the future, we would like to explore parameter tuning for sentence ordering using human evaluation metrics. There are several additional refinements (abstractions) of the extracted sentences which rely on simple post-processing or text cleaning methods which could be performed before sentences are passed to the fusion module. Another interesting direction that we would explore is the possibility of automatically predicting reasonable sentence orderings.

References

Regina Barzilay and Noemie Elhadad. 2002. Inferring strategies for sentence ordering in multidocument news summarization. *Journal of Artificial Intelligence Research*, 17:35–55.

Olivier Bodenreider. 2004. The unified medical language system (umls): integrating biomedical terminology. *Nucleic acids research*, 32(suppl_1):D267–D270.

Jaime Carbonell and Jade Goldstein. 1998. The use of mmr, diversity-based reranking for reordering documents and producing summaries. In *Proceedings of the 21st annual international ACM SIGIR conference on Research and development in information retrieval*, pages 335–336. ACM.

Khyathi Chandu, Aakanksha Naik, Aditya Chandrasekar, Zi Yang, Niloy Gupta, and Eric Nyberg. 2017. Tackling biomedical text summarization: Oaqa at bioasq 5b. *BioNLP 2017*, pages 58–66.

Benoit Favre Dan Gillick and Dilek Hakkani-Tur. 2008. The icsi summarization system at tac 2008. In *In Proceedings of TAC*.

Kevin Donnelly. 2006. Snomed-ct: The advanced terminology and coding system for ehealth. *Studies in health technology and informatics*, 121:279.

Matthew E Falagas, Eleni I Pitsouni, George A Malietzis, and Georgios Pappas. 2008. Comparison of pubmed, scopus, web of science, and google scholar: strengths and weaknesses. *The FASEB journal*, 22(2):338–342.

Katja Filippova. 2010. Multi-sentence compression: Finding shortest paths in word graphs. In *In Proceedings of the 23rd International Conference on Computational Linguistics*, pages 322–330. ACL.

Dan Gillick and Benoit Favre. 2009. A scalable global model for summarization. In *In Proceedings of NAACL*.

Hinrich Schütze, Christopher D Manning, and Prabhakar Raghavan. 2008. *Introduction to information retrieval*, volume 39. Cambridge University Press.

Dan Gillick Taylor Berg-Kirkpatrick and Dan Klein. 2011. Jointly learning to extract and compress. In *In Proceedings of ACL*.

George Tsatsaronis, Georgios Balikas, Prodromos Malakasiotis, Ioannis Partalas, Matthias Zschunke, Michael R Alvers, Dirk Weissenborn, Anastasia Krithara, Sergios Petridis, Dimitris Polychronopoulos, et al. 2015. An overview of the bioasq large-scale biomedical semantic indexing and question answering competition. *BMC bioinformatics*, 16(1):138.

Zi Yang. 2016. *Analytics Meta Learning*. Ph.D. thesis, University of Illinois at Urbana-Champaign.

Zi Yang, Elmer Garduno, Yan Fang, Avner Maiberg, Collin McCormack, and Eric Nyberg. 2013. Building optimal information systems automatically: Configuration space exploration for biomedical information systems. In *Proceedings of the 22nd ACM international conference on Conference on information & knowledge management*, pages 1421–1430. ACM.

Renxian Zhang. 2011. Sentence ordering driven by local and global coherence for summary generation. In *Proceedings of the ACL 2011 Student Session*, pages 6–11. Association for Computational Linguistics.

UNCC QA: A Biomedical Question Answering System

Abhishek Bhandwaldar
Department of Computer Science
UNC Charlotte
abhandwa@uncc.edu

Wlodek Zadrozny
Department of Computer Science
UNC Charlotte
wzadrozn@uncc.edu

Abstract

In this paper, we detail our submission to the BioASQ competition's Biomedical Semantic Question and Answering task. Our system uses extractive summarization techniques to generate answers and has scored highest ROUGE-2 and Rogue-SU4 in all test batch sets.

Our contributions are named-entity based method for answering factoid and list questions, and an extractive summarization techniques for building paragraph-sized summaries, based on lexical chains. Our system got highest ROUGE-2 and ROUGE-SU4 scores for ideal-type answers in all test batch sets.

We also discuss the limitations of the described system, such lack of the evaluation on other criteria (e.g. manual). Also, for factoid- and list -type question our system got low accuracy (which suggests that our algorithm needs to improve in the ranking of entities).

1 Introduction

Most of the recent question answering (QA) systems produce either `factoid` type answers (typically, a phrase or a short sentence) or a summary (typically, returning a few sentences or passages from the text). Creating a natural language answer from relevant passages is still an open problem. Our paper presents is also about providing `factoid` and `summary` answers in BioASQ.

BioASQ is a research competition which is organized by tracks in the biomedical domain. Namely, large-scale online biomedical semantic indexing, biomedical semantic question answering, and information extraction from biomedical literature.

The biomedical QA task is organized in two phases. Phase A deals with retrieval of the relevant document, snippets, concepts, and RDF triples, and phase B deals with exact and `ideal` answer generations. Exact answer generation is required for `factoid`, `list`, and `yes`/`no` type question. `ideal` answer is required for all the question. An `ideal` answer is a paragraph-sized summary of snippets.

BioASQ competition provides the train and test dataset. The training dataset consists of questions, golden standard documents, concepts, and `ideal` answers. The test dataset is split between phase A and phase B. The phase A dataset consists of the questions, unique ids, question types. The phase B dataset consists of the questions, golden standard documents and snippets, unique ids, and question types. Exact answers for `factoid` type questions are evaluated using strict accuracy, lenient accuracy, and MRR (Mean Reciprocal Rank). Answers for the `list` type question are evaluated based on precision, recall, and F-measure. `ideal` answers are evaluated using automatic and manual scores. Automatic evaluation scores consist of ROUGE-2 and ROUGE-SU4 and manual evaluation is done by measuring readability, repetition, recall, and precision.

Summary of our results. In this paper, we present our submission for BioASQ competition. We describe two methods, evaluated on two BioASQ tasks: `ideal` answer and `factoid` type questions. Both methods use conceptual representations based on MetaMap and UMLS.

We compute answers by choosing sentences with the concept chains that are similar to concepts in the question. In `factoid` questions, additionally, our method selects the entities with the highest idf scores.

The first method obtains the best rank for test batch 2,3 and 5 of Phase B of Task 6B. The second method was evaluated on previous year tests with mediocre results.

Proceedings of the 2018 EMNLP Workshop BioASQ: Large-scale Biomedical Semantic Indexing and Question Answering, pages 66–71
Brussels, Belgium, November 1st, 2018. ©2018 Association for Computational Linguistics

2   Related Work

Previous submissions to BioASQ show different approaches by teams taken for answering `ideal`, `factoid`, `list` and `yes/no` questions. OAQA systems(Chandu et al., 2017) use extractive summarization technique for answering `ideal` questions. They have used agglomerative clustering algorithm for similar sentence selection and MMR (Maximal Marginal Relevance) as a sentence similarity measure. Olelo (Neves et al., 2017) proposes a system for getting `yes/no`, `factoid`, `list` and `summary` type question. For summary based questions, the system selects snippets with greatest semantic similarity to question. For `factoid` and `list` type questions, they select an answer, based on matching predicates, and for `yes/no` question, they do sentiment analysis. (Aliod, 2017) have submitted the system for `ideal` answers only. They propose an extractive summarization approach that does sentence segmentation, ranks the sentences based using a scoring function, and return the top n sentences as the answer.

(Sarrouti and Alaoui, 2017) describes a system which retrieves snippets from relevant documents, re-ranks using BM25 model and finally concatenates top two snippets. For `factoid` and `list` type, their system extracts biomedical entities from relevant snippets, ranks them based on their frequency, and return top n. For `yes/no` they use sentiment analysis.

(Wiese et al., 2017) proposes a deep learning based approach to answering the `factoid` and `list` type question. The system is based on FastQA (Weissenborn et al., 2017), which is trained on SQUAD dataset and fine-tuned on BioASQ dataset to select a substring in relevant snippets as the final answer.

Other non-BioASQ systems that are capable of question answering include IMB's Watson (Ferrucci, 2012). The Watson system is an open domain question answering system that won the TV game-show Jeopardy! in 2011. The system worked by pipelining different components like question decomposition, hypothesis generation, hypothesis and evidence scoring, and answer generation. A more recent approach using deep learning is dynamic memory networks(Kumar et al., 2015). It uses the SQUAD dataset and simulates episodic memory using recurrent neural networks; it can also answer questions that require transitive reasoning. The SQUAD dataset is a reading comprehension dataset that requires the system to find a segment of text as the answer for a given question. Most of the systems based on SQUAD dataset are factoid answering system, and do not generate natural language answers.

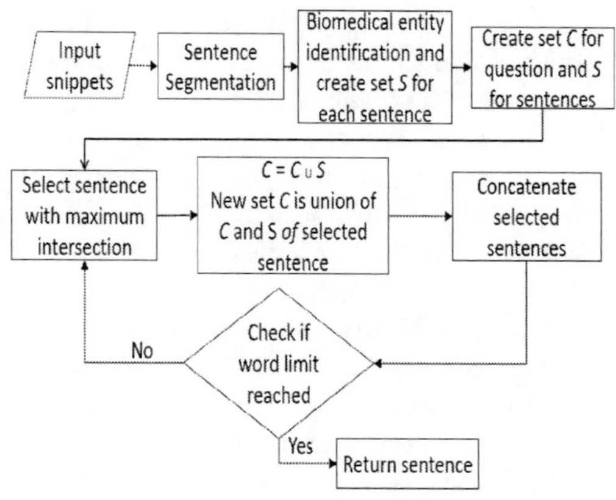

Figure 1:   Our `summary` type question pipeline. The input is a list of snippets and the biomedical entity identification is done using MetaMap and UMLS.

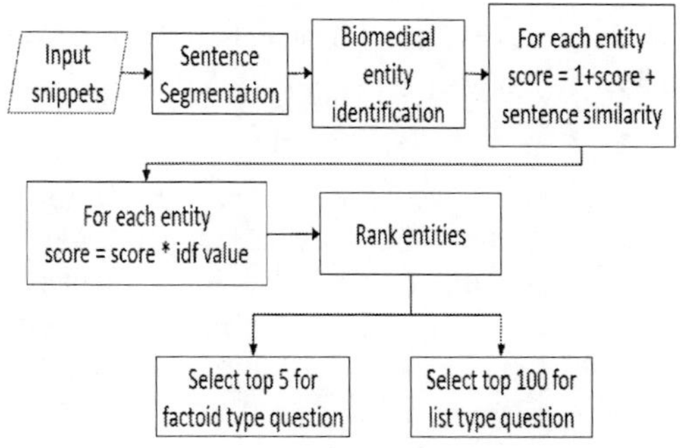

Figure 2:   `factoid` and `list` type question pipeline. The input is a list of snippets and the biomedical entity identification is done using MetaMap and UMLS. The entities are scored using frequency, idf, and sentence similarity score.

3   Our Question Answering Pipeline

3.1   Ideal Answer

For `ideal` question type, we use extractive summarization to generate the answer. The word limit of the `ideal` answer is 200 words. Our extractive summarization pipeline is inspired by lexical chaining.

3.1.1 Lexical Chaining

Lexical chaining is a technique for identifying semantically related words that represent the concept or the semantic meaning of a sentence. A lexical chain does not describe the grammatical structure of a sentence. This technique has been used for text summarization by ranking sentences with similar ideas. Top sentences are then combined to produce a final summary. They are also helpful in word sense disambiguation (Okumura and Honda, 1994) by providing context to a term and capturing concept represented by that term. (Xiong et al., 2013) describes a lexical chain based approach for machine translation. Implementations of the lexical chain approach vary based on their applications. For example, (Reeve et al., 2006) describes the use of lexical chains for biomedical document summarization. Their technique uses a chain of concepts found by mapping biomedical terms to concepts using UMLS(Unified Medical Language System). UMLS is a meta thesaurus for biomedical terms. Once the concepts are found the strongest chain is found by sorting the chains based on a scoring function that takes into account various factors like word frequency, distinct concepts, word distance, and homogeneity. Finally, top sentences are used to generate the final summary.

3.1.2 Our Approach

For `ideal` answers, we use extractive summarization technique on relevant snippets. Our extractive summarization pipeline uses lexical chaining for sentence similarity and ranking. We then select the top N sentences such that the total number of words doesn't exceed the 200-word limit, and concatenate them to form the summary or the final answer. In our algorithm, we first do sentence segmentation on relevant snippets, and pass each sentence through the MetaMap tool. The MetaMap tool identifies all biomedical entities contained in the statement and returns the preferred name and semantic type for every biomedical entity.

For every sentence, we create a set C containing the semantic types of all biomedical terms in the sentence. We also create a similar set S for the question text. Next, we find the intersection of set C of every sentence with the question set S and assign a score as the number of intersecting terms. We select the sentence with a maximum score and add it to the summary list. We also aug-

ment the question set S by doing the union of set C of the selected sentence with set S. We then use the new set S to find intersection with set C of remaining sentences. We repeat this procedure until we reach 200-word limit. Finally, to generate the summary, we concatenate the list of selected sentences to create the final answer.

In terms of tools, we used Stanford CoreNLP for snippet segmentation to get sentences. We use custom code using the Java API to MetaMap to get the concepts. MetaMap is also responsible for tokenizing, word sense disambiguation, connecting to UMLS and getting all the required mappings.

3.2 `factoid` and `list` type answers

For `factoid` type questions we are required to return a list of 5 entity names. The `list` type questions need to return a list of at most 100 entity names each of no more than 100 characters.

For answering the `factoid` type question, we use a similar technique as the `summary` generation pipeline, with additional scoring factors, and scoring at entity level, rather than sentence level. For each sentence, we get a list of biomedical entities using MetaMap. We score each entity using a scoring function that uses the entity frequency, the idf weight (the inverse document frequency of that entity), and the sentence similarity score found by the intersection of semantic set C of that sentence and question set S. Finally, we rank the list based on the scores and select top 5 entities as the answer for the `factoid` type question, and top 100 for the `list` type. We use the idf scores to eliminate common biomedical words or phrases. To get the idf score we downloaded and indexed Annual Baseline Medline repository, PubMed, using Lucene. We then use the Lucene indexes to get the term frequency and document count to calculate the idf score. For a multi-word entity, the idf score is the maximum of the idf scores of the individual tokenized words. This way a biomedical entity with even a single rare word will be ranked higher.

4 Results

We submitted results of our system for Phase B of task 6B. Phase B consisted of 5 test batches. We submitted our results for test batches 2,3 and 5 for `ideal` answers. We also evaluated our system on an older test batch sets using the BioASQ oracle. `ideal` answers are manually assessed using readability, repetition, recall, and precision and automatically by using ROUGE-2 and Rogue-SU4 scores. At the time of submission we did not have the manual scores; hence we only report the automatic scores. Table 1 and 5 shows our results on previous years dataset. As can be seen from Table 2 our system gave highest ROUGE-2 and ROUGE-SU4 scores among all systems on every test batch set.

Test Batch	ROUGE-2	ROUGE-SU4
Task 5B Batch 5	0.7188	0.7062
Task 5B Batch 4	0.7363	0.7258
Task 5B Batch 3	0.7802	0.7769
Task 5B Batch 2	0.6918	0.6903
Task 5B Batch 1	0.6716	0.6712
Task 4B Batch 5	0.7266	0.7250
Task 4B Batch 4	0.7196	0.7177
Task 4B Batch 3	0.6364	0.6527
Task 4B Batch 2	0.6777	0.6897
Task 4B Batch 1	0.6918	0.7024

Table 1: Results of `ideal` answers on task 4B and 5B test batch sets using BioASQ oracle. The results are arranged from most recent to least. The table shows ROUGE-2 and ROUGE-SU4 scores.

For BioASQ task 6B, we submitted `ideal` answers with summary created by selecting the only top sentence and with 200 word limit. The system that created the summary with 200 word limit gave highest ROUGE-2 and ROUGE-SU4 scores on every test batch set (2,3, and 5) that we submitted. Table 2 details this results.

Test Batch	Test Batch	ROUGE-2	ROUGE-SU4
UNCC System 1	Task 6B Batch 2	0.5833	0.6015
UNCC System 1	Task 6B Batch 3	0.6184	0.6290
UNCC System 2	Task 6B Batch 3	0.1973	0.1947
UNCC System 1	Task 6B Batch 5	0.7250	0.7122
UNCC System 2	Task 6B Batch 5	0.3846	0.3759

Table 2: Results of `ideal` answers on task 6B test batch sets using BioASQ oracle. The table shows ROUGE-2 and ROUGE-SU4 scores of UNCC System 1 and 2. UNCC System 1 submitted the summary created with 200 word limit. UNCC System 2 submitted summary created by selecting only top sentence.

For `factoid` and `list` type questions we did not submit results for task 6B, and we only report results from previous year's test batch sets using BioASQ oracle. Table 3 shows scores of

Test Batch	Factoid SAcc	Factoid LAcc	Factoid MRR
Task 5B Test Batch 1	0.1200	0.1600	0.1333
Task 5B Test Batch 2	0.0323	0.1290	0.0575
Task 5B Test Batch 3	0.0385	0.0769	0.0462
Task 5B Test Batch 4	0.0303	0.0909	0.0455
Task 5B Test Batch 5	0.0571	0.1429	0.0786

Table 3: Result of `factoid` type question on task 5B of BioASQ. The scores include the Lenient accuracy LAcc, strict accuracy SAcc, and MRR(Mean Reciprocal Rank).

`factoid` type question and Table 4 shows scores of `list` type questions.

Test Batch	List Mean Precision	List Recall	List F-measure
Task 5B Test Batch 1	0.0241	0.3252	0.0441
Task 5B Test Batch 2	0.0353	0.2700	0.0600
Task 5B Test Batch 3	0.0195	0.3673	0.0367
Task 5B Test Batch 4	0.0250	0.2051	0.0389
Task 5B Test Batch 5	0.0391	0.2867	0.0630

Table 4: Result of `list` type question on task 5B of BioASQ. The results are evaluated on Mean Precision, Recall, and F-measure.

Test Batch	ROUGE-2	ROUGE-SU4
Task 3B Batch 5	0.5651	0.5672
Task 3B Batch 4	0.5848	0.5950
Task 3B Batch 3	0.5994	0.6128
Task 3B Batch 2	0.5451	0.5674
Task 3B Batch 1	0.5240	0.5368
Task 2B Batch 5	0.3967	0.4180
Task 2B Batch 4	0.4201	0.4458
Task 2B Batch 3	0.4731	0.4754
Task 2B Batch 2	0.4075	0.4258
Task 2B Batch 1	0.5313	0.5326
Task 1B Batch 2	0.3319	0.3596
Task 1B Batch 1	0.3032	0.3276

Table 5: Results of `ideal` answers on task 3B, 2B and 1B test batch sets using BioASQ oracle. The results are arranged from most recent to least. The table shows ROUGE-2 and ROUGE-SU4 scores.

5 Discussion

In this section we discuss the limitations of our work and address the reviewers comments not included in the revision of previous sections.

Our system was only evaluated by the ROUGE scores. However, high ROUGE results do not always imply good manual scores. In our approach we use the sentences of the passages without any changes. Thus, as observed by one of the reviewers, the manual score Repetition might be high; neither it is clear what is the impact of this approach on Readability. Furthermore, high ROUGE scores could be a side effect of a situation when the total number of words in all passages is less than 200.

Our evaluation was only done on ROUGE, because while building the system we only had access to old batch sets, and it was our first attempt to participate in Phase B of this competition. This said, our system tries to find the best candidate answers, and then concatenates them. So, further work needs to be done to convert the information from candidate snippets to a natural language answer that makes sense, and does not include any irrelevant information. We hope to address in the next year competition.

Regarding Repetition, our system first does sentence segmentation to get a list of snippets. Sometimes the snippets are overlapping and can have common sentences. Our system takes care of not repeating these sentences. What the system lacks is detecting sentences that are semantically similar and only consider one of them. Again – future work.

Regarding Readability, our concatenation is done such that each concatenated sentence is separated by period, hence usually making a coherent passage. Still the sentences might not follow a particular flow, and this might affect the readability score.

Another issue worth discussing is our approach to scoring the candidate answers. First, while scoring on the basis of term frequency is common, we use it (like other systems do), but we combine it with a `summary` pipeline score and the idf score. Second, we would have gone for machine learning techniques, but we felt we did not have enough labeled data.

One can argue that MetaMap doesn't always capture the all biomedical entities. However, we didn't face this problem. Although expanding candidate answers to include noun phrases could possibly improve the recall in generating candidate answers.

The OAQA system (Chandu et al., 2017) uses extractive summarization techniques like our system and the difference lies in sentence similarity. Our extractive summarization algorithm also shares similarity with Maximal Margin Relevance (MMR) in that both get sentence relevance score by comparing with question and other selected sentences. Our extractive summarization technique gives us higher ROUGE score than OAQA. Olelo system (Neves et al., 2017)and our system have similar pipeline in generation of summary and the only difference is in the way we do sentence similarity.

For `factoid` and `list` type questions our system does not perform well and the system can be improved by introducing better ranking algorithm, improved entity identification and filtering (at this time we use idf score to find out very common entities), and better relevance score between entity and the question.

6 Conclusion

In this paper, we showed our system's extractive summarization technique using lexical chains, or, more accurately, conceptual chains). We introduced an extractive summarization techniques for building paragraph-sized summaries. We have seen that use of the set of semantic type has proved very capable in ranking candidate answer sentences. Our system got highest ROUGE-2 and ROUGE-SU4 scores for `ideal` answers in all test batch sets.

We also showed a method to answer factoid and `list` type question. For these type of questions our system got low accuracy, which suggests that our algorithm needs to improve in ranking the entities. We plan to address these and other issues in future experiments.

Acknowledgment. We would like to thank the referees for their comments and suggestions. All the remaining faults are ours.

References

Diego Mollá Aliod. 2017. Macquarie university at bioasq 5b - query-based summarisation techniques for selecting the ideal answers. In (Cohen et al., 2017), pages 67–75.

Khyathi Chandu, Aakanksha Naik, Aditya Chandrasekar, Zi Yang, Niloy Gupta, and Eric Nyberg. 2017. Tackling biomedical text summarization: OAQA at bioasq 5b. In (Cohen et al., 2017), pages 58–66.

Kevin Bretonnel Cohen, Dina Demner-Fushman, Sophia Ananiadou, and Junichi Tsujii, editors. 2017. *BioNLP 2017, Vancouver, Canada, August 4, 2017.* Association for Computational Linguistics.

D.A. Ferrucci. 2012. Introduction to "this is watson". 56:1:1–1:15.

Ankit Kumar, Ozan Irsoy, Jonathan Su, James Bradbury, Robert English, Brian Pierce, Peter Ondruska, Ishaan Gulrajani, and Richard Socher. 2015. Ask me anything: Dynamic memory networks for natural language processing. *CoRR*, abs/1506.07285.

Mariana L. Neves, Fabian Eckert, Hendrik Folkerts, and Matthias Uflacker. 2017. Assessing the performance of olelo, a real-time biomedical question answering application. In (Cohen et al., 2017), pages 342–350.

Manabu Okumura and Takeo Honda. 1994. Word sense disambiguation and text segmentation based on lexical cohesion. In *15th International Conference on Computational Linguistics, COLING 1994, Kyoto, Japan, August 5-9, 1994*, pages 755–761.

Lawrence H. Reeve, Hyoil Han, and Ari D. Brooks. 2006. Biochain: lexical chaining methods for biomedical text summarization. In *Proceedings of the 2006 ACM Symposium on Applied Computing (SAC), Dijon, France, April 23-27, 2006*, pages 180–184. ACM.

Mourad Sarrouti and Said Ouatik El Alaoui. 2017. A biomedical question answering system in bioasq 2017. In (Cohen et al., 2017), pages 296–301.

Dirk Weissenborn, Georg Wiese, and Laura Seiffe. 2017. Making neural QA as simple as possible but not simpler. In *Proceedings of the 21st Conference on Computational Natural Language Learning (CoNLL 2017), Vancouver, Canada, August 3-4, 2017*, pages 271–280. Association for Computational Linguistics.

Georg Wiese, Dirk Weissenborn, and Mariana L. Neves. 2017. Neural question answering at bioasq 5b. In (Cohen et al., 2017), pages 76–79.

Deyi Xiong, Yang Ding, Min Zhang, and Chew Lim Tan. 2013. Lexical chain based cohesion models for document-level statistical machine translation. In *Proceedings of the 2013 Conference on Empirical Methods in Natural Language Processing, EMNLP 2013, 18-21 October 2013, Grand Hyatt Seattle, Seattle, Washington, USA, A meeting of SIGDAT, a Special Interest Group of the ACL*, pages 1563–1573. ACL.

An Adaption of BIOASQ Question Answering dataset for Machine Reading systems by Manual Annotations of Answer Spans

Sanjay Kamath
LIMSI, LRI
Univ. Paris-Sud, CNRS
Université Paris-Saclay
Orsay, France

Brigitte Grau
LIMSI, CNRS
ENSIIE
Université Paris-Saclay
Orsay, France

Yue Ma
LRI
Univ. Paris-Sud, CNRS
Université Paris-Saclay
Orsay, France

sanjay@lri.fr   brigitte.grau@limsi.fr   yue.ma@lri.fr

Abstract

BIOASQ Task B Phase B challenge focuses on extracting answers from snippets for a given question. The dataset provided by the organizers contains answers, but not all their variants. Henceforth a manual annotation was performed to extract all forms of correct answers. This article shows the impact of using all occurrences of correct answers for training on the evaluation scores which are improved significantly.

1 Introduction

BIOASQ[1] challenge is a large-scale biomedical semantic indexing and question answering task (Tsatsaronis et al., 2015) which has been successful for 5 years. The challenge proposes several tasks using Biomedical data. One of the tasks focuses on Biomedical question answering (Task B Phase B - we further refer it as B) where the goal is to extract answers for a given question from relevant snippets.

Several teams have participated actively, and a noticeable aspect is that the results of the task B are much lower compared to open domain QA evaluations, as in SQUAD[2]. Some reasons can be the low dataset size and the format of the answers provided by the organizers. Bioasq provides only certain answer forms in the gold standard data and not all the variants of the answers in the given snippets.

In this paper, we study the influence of enriching the training data by manually annotated variants of gold standard answers on the evaluation performance. We show the impact of the enriched data by experimenting on 5B and 6B training datasets. Our method outperforms the best-

performing systems from Bioasq 5B by 7.3% on strict accuracy and 18% on lenient accuracy.

2 Related Work

Several works in the past BIOASQ tasks have used classical question answering pipeline architecture adapted to the biomedical domain. Some use the domain-specific information from UMLS tools such as Metamap (Schulze et al., 2016), along with other NLP tools like Corenlp, LingPipe (Yang et al., 2016). A typical question answering pipeline consists of:

1. Question processing for question type detection and lexical answer type detection.

2. Document retrieval (Task B Phase A)

3. Answer extraction by answer re-ranking on the candidate answers generated in the previous phases, done in a supervised learning manner.

In the open domain, deep learning models are extensively used in machine reading task. Datasets such as MS Marco by (Nguyen et al., 2016), SQUAD by (Rajpurkar et al., 2016) and Wikireading by (Hewlett et al., 2016) have made it easier for deep learning models to perform better on machine reading task. One of the first attempts to use deep learning algorithms for the Bioasq task was reported in BIOASQ 5 by (Wiese et al., 2017b) where the dataset was adapted to be used as a machine reading dataset whose goal is to extract answers from snippets. The authors use a model trained on open domain questions, and perform domain adaptation to biomedical domain using BIOASQ data. Their system got one of the best results whose methods are reported in the section 5.

---

[1] http://bioasq.org/
[2] https://rajpurkar.github.io/SQuAD-explorer/

Proceedings of the 2018 EMNLP Workshop BioASQ: Large-scale Biomedical Semantic Indexing and Question Answering, pages 72–78
Brussels, Belgium, November 1st, 2018. ©2018 Association for Computational Linguistics

3   Evaluation and training data

Bioasq 6 is the sixth challenge and the evaluation measures for Bioasq task B has always been the same. Strict Accuracy, Lenient Accuracy and Mean Reciprocal Rank (MRR) are the 3 evaluation measures used. To compute the scores, the exact match of strings between the predictions and the gold standard answers is used to decide if a system answer is correct. Strict accuracy is the rate of top 1 exact answers. Lenient accuracy is the rate of exact answers in top 5 predictions. MRR is the mean reciprocal rank computed on the top 5 system answers. These measures have been the same since the 1st challenge, although the first four challenges had triples and concepts along with snippets in the data. In the last two challenges, only relevant snippets for questions are released.

Similar evaluations are performed in machine reading tasks like in SQUAD where top 1 accuracy and F1 scores are computed by comparing exact matching strings. One main assumption in machine reading task is that the answer strings are substrings of the snippets, which implies that answers have to be extracted from the snippets.

In Bioasq, the answers are curated by human experts by analyzing the triples, concepts, and snippets (or paragraphs). Thus, the Bioasq dataset and evaluation measures are very similar to that of machine reading task, but the major difference apart from the dataset size are the answers instances provided as gold standard which does not contain all the occurrences, abbreviations, different forms of answers which are present in the snippets.

In (Wiese et al., 2017b), the authors transform Bioasq Phase B as a machine reading task with domain adaptation. Gold standard answer strings and their offsets are automatically searched in the snippets for exact match and treated as answers if only they are found in the snippets, i.e., the answer string must be a substring of the snippet. By doing so the dataset size is reduced to 65% of Bioasq 5 train set which was suitable for adaptation. Other 35% of the questions did not have matching answers in the snippets, because of different variants of answers in the snippets, missing abbreviations, or irrelevant snippets.

This snippet annotation method can result in:

- False positive: an answer mentioned in the snippet which does not answer the question.

- False negative: a snippet answers the question but does not have the exact string compared to the gold standard string.

We found that in Bioasq 6B training dataset for factoid questions, 205 out of 619 questions have false negative answers (33% of the dataset) which may result in some problems:

- Less data for learning;

- The model does not learn to extract all the variants;

- Evaluation is done using such gold standard data which will lower the results even though the model is performing well.

Below are some examples for which the answers returned from a reference system is correct (when evaluated manually) but the automatic evaluation classifies it as incorrect.

> **Q: Which calcium channels does ethosuximide target?**
> **P: ..neuropathic pain is blocked by ethosuximide, known to block T-type calcium channels,..**
>
> **Prediction:** T-type calcium
> **Gold standard:** T-type calcium **channels**

Example 1: Missing keywords

> **Q: Which disease can be treated with Delamanid?**
> **P: In conclusion, delamanid is a useful addition to the treatment options currently available for patients with MDR-TB.**
>
> **Prediction: MDR-TB**
> **Gold standard: tuberculosis**

Example 2: Abbreviations

In example 1, because of a missing word "channels", the predicted answer is marked incorrect. In example 2, MDR-TB stands for *Multi-drug-resistant tuberculosis*, which is from a relevant snippet but since the gold standard has only *tuberculosis*, it is marked incorrect. Contextually both are valid answers.

To overcome this problem and enrich the answer space correctly, we manually annotated 618 factoid question-answers pairs from training dataset of 6B task, by annotating the substring of the gold standard answers in the snippets, and adding answers with abbreviations, multi-word answers, synonyms, that are likely correct answers. We explain this in detail in the following section.

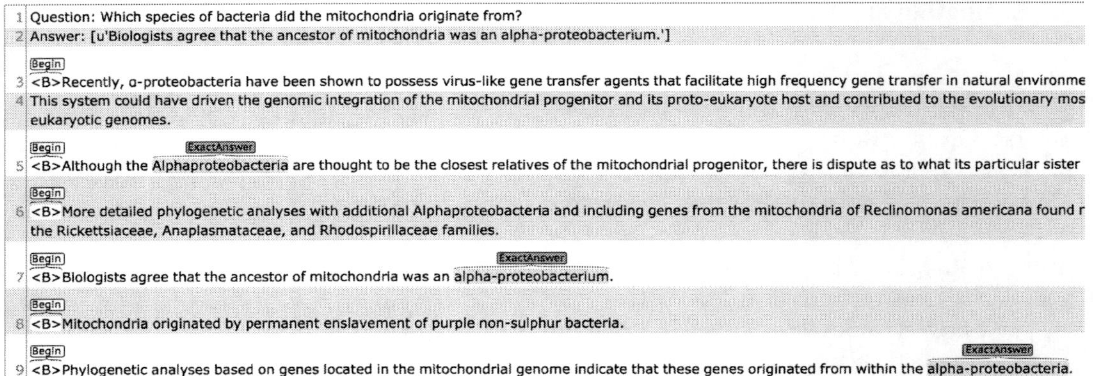

Figure 1: Brat annotation tool

4 Annotations

This section presents the details of the annotations performed manually on the BIOASQ 6B training dataset and presents some statistics.

Our annotations include the following type of answers:

- Exact Answer - Exact match with gold standard (GS) answers, which can also be annotated automatically, and different variants of the answers. For example, the annotation of a single GS answer *"Transcription factor EB (TFEB)"* resulted in 3 annotations, *"Transcription factor EB"*, *"TFEB"*, *"Transcription factor EB (TFEB)"*.

- Lenient Answer - a more general form or a more specific form of an answer. An example is *"Telomerase"* for *"Human telomerase reverse transcriptase"*.

- Paragraph Answer - The answer matches with gold standard but the snippet alone is not relevant to the question.

We came across several kinds of snippets. A supporting snippet, or answering snippet, is a snippet that contains the answer and enough elements for justifying it. It is a correct answer to the question (snippet starting at line 5 in Figure 1 for example). A snippet that contains the answer without justification towards the question will not be annotated with the answer as correct and is considered as a non-supporting snippet (snippet starting at line 3 in Figure 1). A snippet that does not contain the answer cannot be a supporting snippet, henceforth it is an irrelevant snippet (snippet starting at line 8 in Figure 1).

We use Brat[3] annotation tool by (Stenetorp et al., 2012) shown in Fig. 1 to perform the manual annotations of the snippets with the answer to the question. The annotations done include the answer string along with their character offsets in the snippet. Answers were annotated by 3 people from computer science background and multiple discussions were held to discuss problematic answers which involved looking upon the internet for some medical term meanings.

Annotations were initially done on the Bioasq 5B training set and the additional questions from 5B test sets whose answers are present in the 6B training set were annotated later on 6B data. So the changes done (if any) on 6B training set for previous year questions from 5B set are not considered.

The annotation files are freely available[4] and can be used by researchers who can get the Bioasq dataset.

Some statistics of the dataset are listed in Table 1 for the automatically annotated answers from gold standard data and the fully annotated data with manual annotations. The annotations are done on 618 BIOASQ 6B training dataset questions. Out of 619 factoid questions, 1 question does not have any snippets.

Only 426 questions contain answers from automatic annotation.

"Answers" are the count of answers present in the snippets. *Avg* score represents an average over the total number of questions (i.e. 618). Since in

---

[3]http://brat.nlplab.org
[4]https://zenodo.org/record/1346193#.W3_WUZMzZQI

Count	Gold std. annotations		Full annotations	
	Avg	Total	Avg	Total
Answers	0.8	500	2.9	1814
Snippets	7.7	3286	8	4965
Questions	-	426	-	618

Table 1: Annotation statistics

gold standard data, only 426 questions have gold answers in snippets, it is normal for the average to fall below 1. It is clear from the table that the full annotated data contain at least 3 times (1814 answers) more the number of candidate answers over the provided gold standard ones (500 answers).

We found that some answers contained the whole snippet as an answer and that 3503 snippets are repeated in the 6B train set. After filtering those repeated snippets we found 3286 different snippets containing exact matching answers extracted automatically from gold standard data and 4965 unique snippets manually annotated with correct answers.

5   Experiments

The goal of our experiments is to study the impact of the data augmentation on training and evaluating a system.

Henceforth, we follow the process of (Wiese et al., 2017b) and use a machine reading model developed by (Chen et al., 2017) that is pre-trained on SQUAD dataset (Rajpurkar et al., 2016) for open domain questions and fine tuned to biomedical questions.

To study the impact on the training process and the evaluations, we train the models using separately the automatically annotated data and the fully manually annotated data. We also evaluate them using both kinds of data separately.

5.1   QA system overview

We present here the adaptation of an existing model named DRQA reader by (Chen et al., 2017) to the biomedical domain as presented in (Kamath et al., 2018).

DRQA reader has three components:

1. Input layer: where the input question words and input passage words are encoded using a pretrained word embedding space.

2. Neural layer: RNN or LSTM networks.

3. Output layer or decoding layer: where the outputs are start and end tokens representing a span of an extracted answer.

The reader model takes as input, the question sentence and the answering snippet and predicts the substring of the snippet that is the answer.

In the input layer, word embeddings are used to encode the words of snippets and questions into vectors, along with textual features such as Part of Speech tags, Named-Entity tokens, Term frequencies of the words in the snippet. The authors use *aligned question embeddings* where an attention score captures the similarity between snippet words and questions words. The neural layer, where the core DNN model is defined, uses different NN architectures to capture semantic similarities between the question/snippet pairs. It use LSTMs to encode the snippets and an RNN to encode the questions. In the output layer, two independent classifiers use a bilinear term to capture the similarity between snippet words and question words and compute the probabilities of each token being start and end of the answer span. We take all possible scores of start and end token predictions and restrict the span between start and end tokens to 15 tokens. We perform an outer product between these scores and consider top 5 spans using an argmax value to get these final predictions.

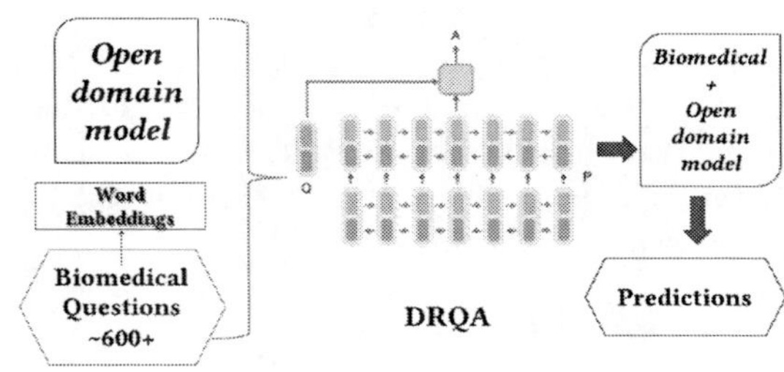

Figure 2:   Transfer learning from open domain to biomedical domain

Domain adaptation (also referred to as fine tuning) is performed on the BIOASQ dataset as shown in Figure 2 where the model is pre-trained with SQUAD dataset and fine-tuned with BIOASQ before predicting on test sets. Pre-training is training a model from scratch with randomly initialized weights. Fine-tuning is training on a model with previously trained weights rather than randomly initialized ones. The advantage of pre-

Train set		5B				6B			
Finetune		Gold		Anno.		Gold		Anno.	
Eval	DeepQA	Gold	Anno.	Gold	Anno.	Gold	Anno.	Gold	Anno.
Strict	-	0.2551	**0.2962**	0.1666	**0.3333**	0.2669	**0.3090**	0.2265	**0.3948**
Lenient	-	0.4156	**0.4444**	0.2991	**0.5843**	0.4417	**0.4724**	0.3511	**0.6197**
MRR	0.2620	0.3138	**0.3425**	0.2148	**0.4322**	0.3334	**0.3718**	0.2728	**0.4765**

Table 2: K-fold evaluation on different train sets with *Gold* and *Anno* data. DeepQA scores are presented by (Wiese et al., 2017a)

training with SQUAD dataset is that the DNN model will learn and perform better while trained on a larger training dataset. Since the target dataset is in the biomedical domain, finetuning the previously learned model will have a positive impact on the test set predictions, as shown in (Wiese et al., 2017a).

Several embedding spaces were tested as input vectors (Kamath et al., 2017) and the best performing ones which were the Glove embeddings trained on common crawl data with 840B tokens, were chosen as input to the system. Unknown words were initialized as zero vectors.

As BIOASQ questions have several answering snippets, we treat each question and a snippet as a training sample which might often result in repeated questions with different snippets, i.e. for each training example, there is a question, a unique snippet and the start and end token string offsets of the answer in the snippet. Our model predicts one scored answer per snippet, and the final result is made of the ordered list of answers for the same question. We consider only the top 5 answers.

5.2 Datasets

We perform fine-tuning on two datasets namely

- BIOASQ 5B training set, which contains the 4B training data + the answers of the 4B test data - We term it as 5B.

- BIOASQ 6B training set, which contains the 5B training data + the answers of the 5B test set - We term it as 6B.

We term the automatically annotated training data as *Gold*, and manually annotated training data as *Anno*.

The pre-trained model on open domain QA data is fine-tuned on the above listed Bioasq datasets separately. Evaluation is performed by K-fold

cross validation because of the small scale of the data (Table 2), and on the official test sets of Bioasq 5B (Table 3), which were separated from the training data while fine-tuning.

The explanation of scores reported in table 2 and 3 along with the corresponding experiments on the datasets listed above, is as follows. On the data of *Trainset* mentioned in the first row, we fine-tune it with *Finetune* data on the second row - which is *Gold* or *Anno.* version of the answers.

The official evaluation measures[5] using *Gold* or *Anno.* version of the answers are highlighted in the third row. The strict and lenient accuracies along with the MRR are reported.

Gold version of 5B data contains 313 questions and *Gold* version of 6B data contains 428 questions. We consider the remaining questions with no matching answers as incorrectly answered, hence evaluating over all the questions of the datasets (5B - 486 questions, 6B - 618 questions). Annotated 5B data contains 483 questions and 6B data contains 618 questions.

Overall results of 5B test sets presented in Table 3 are evaluated on 150 questions from the test sets of 5B challenge whose gold standard answers are present in 6B challenge train set.

To compare our scores with the ones reported in (Wiese et al., 2017a) and also since the size of the dataset is small, we perform K-Fold (5) evaluations which are reported in Table 2. To compare with previously reported official test scores in Bioasq 5, we train on 5B training set and test on 5B test sets which are reported in Table 3.

6 Results

The results shown in table 2, 3 and 4 highlights the improvements using manually annotated data over the automatically annotated data on the QA performance as well as the evaluations with *Gold*

---

[5]https://github.com/BioASQ/Evaluation-Measures

Train set	5B					
Finetune			Gold		Anno.	
Eval	(Wiese et al., 2017b)	Lab Zhu, Fudan Univer	Gold	Anno.	Gold	Anno.
Strict	0.3466	0.3533	0.3533	**0.42**	0.3133	**0.4266**
Lenient	0.5066	0.4533	0.54	**0.64**	0.5	**0.6866**
MRR	-	-	0.4256	**0.5042**	0.3884	**0.5258**

Table 3: Overall results calculated on official test sets from 5B task. Scores from (Wiese et al., 2017b) and *Lab Zhu, Fudan Univer* are reported in Bioasq 5.

Train set	5B					
Finetune			Gold		Anno.	
Eval	(Wiese et al., 2017b)	Lab Zhu, Fudan Univer	Gold	Anno.	Gold	Anno.
Batch 1	0.5600	0.4200	0.4733	**0.5733**	0.4933	**0.6066**
Batch 2	0.4086	0.4839	0.4274	**0.5510**	0.3387	**0.5215**
Batch 3	0.4308	0.3846	0.4070	**0.4198**	0.3185	**0.3955**
Batch 4	0.3025	0.2601	0.3595	**0.4474**	0.4444	**0.6196**
Batch 5	0.3924	0.4524	0.4271	**0.4771**	0.3452	**0.5023**

Table 4: MRR results calculated batchwise on 5B official test sets.

and *Anno.* versions of answers.

In Table 2, training on *Gold* and evaluating on *Gold* are the baseline scores. *DeepQA* MRR score is the K-fold evaluation score of MRR reported on 5B train set by (Wiese et al., 2017a).

Comparing the *DeepQA* MRR score with the *Gold* and *Anno.* 5B versions, there is an improvement of at least 17% (*Anno.* training and *Anno.* evaluation) to 8% (*Gold* training and *Anno.* evaluation).

In terms of accuracy, training the model on *Anno.* version and evaluating on *Anno.* version of answers fetch best results by 3.68% and 8.58% on Strict accuracy, 14% and 14.73% on Lenient accuracy in 5B and 6B respectively.

Training on *Anno.* and evaluating on *Gold* has low scores in almost all experiments because of the model which learns on different forms of answers, therefore predicts different forms of answers which are not present in the *Gold* version.

In Table 3, because of a low number of questions in the official test sets ranging from 25 to 35 questions, the scores are computed over all 5B batch test sets by using individual batch results for the number of correct answers from official Bioasq scores and calculating the score over a total number of questions in the 5 batches (5B test sets - 150 questions). The scores by (Wiese et al., 2017b) and *Lab Zhu, Fudan Univer* are the best official results in Bioasq 5. We calculated strict and le-

nient accuracy as mentioned above and our scores are better than both best official results by 6.67% for strict accuracy and 13.34% lenient accuracy on *Gold* version training, 7.33% for strict accuracy and 18% lenient accuracy on *Anno.* version training.

In Table 4, MRR scores are reported separately for each batch. MRR scores in general have the best scores compared to both (Wiese et al., 2017b) and *Lab Zhu, Fudan Univer* by training on *Anno.* and evaluating on *Anno.* versions.

7 Conclusion and Future Work

We present the importance of using all variants of answers in the snippets for adapting the Bioasq dataset to machine reading task format. We show that the results can be much higher than the officially reported ones if all the variants of the answers are annotated correctly in the training sets. We perform manual annotations to show this impact. Future work would focus on automatic detection of these variants of answers in the snippets.

References

Danqi Chen, Adam Fisch, Jason Weston, and Antoine Bordes. 2017. Reading wikipedia to answer open-domain questions. In *Proceedings of ACL 2017*, pages 1870–1879.

Daniel Hewlett, Alexandre Lacoste, Llion Jones, Illia Polosukhin, Andrew Fandrianto, Jay Han, Matthew

Kelcey, and David Berthelot. 2016. Wikireading: A novel large-scale language understanding task over wikipedia. *arXiv preprint arXiv:1608.03542.*

Sanjay Kamath, Brigitte Grau, and Yue Ma. 2017. A study of word embeddings for biomedical question answering. In *SIIM'17.*

Sanjay Kamath, Brigitte Grau, and Yue Ma. 2018. Verification of the expected answer type for biomedical question answering. In *Companion Proceedings of the The Web Conference 2018*, WWW '18, pages 1093–1097, Republic and Canton of Geneva, Switzerland. International World Wide Web Conferences Steering Committee.

Tri Nguyen, Mir Rosenberg, Xia Song, Jianfeng Gao, Saurabh Tiwary, Rangan Majumder, and Li Deng. 2016. Ms marco: A human generated machine reading comprehension dataset. *arXiv preprint arXiv:1611.09268.*

Pranav Rajpurkar, Jian Zhang, Konstantin Lopyrev, and Percy Liang. 2016. Squad: 100,000+ questions for machine comprehension of text. *arXiv preprint arXiv:1606.05250.*

Frederik Schulze, Ricarda Schüler, Tim Draeger, Daniel Dummer, Alexander Ernst, Pedro Flemming, Cindy Perscheid, and Mariana Neves. 2016. Hpi question answering system in bioasq 2016. In *Proceedings of the Fourth BioASQ workshop*, pages 38–44.

Pontus Stenetorp, Sampo Pyysalo, Goran Topić, Tomoko Ohta, Sophia Ananiadou, and Jun'ichi Tsujii. 2012. Brat: a web-based tool for nlp-assisted text annotation. In *Proceedings of the Demonstrations at the 13th Conference of the European Chapter of the Association for Computational Linguistics*, pages 102–107. Association for Computational Linguistics.

George Tsatsaronis, Georgios Balikas, Prodromos Malakasiotis, et al. 2015. An overview of the bioasq large-scale biomedical semantic indexing and question answering competition. *BMC Bioinformatics*, 16(1):138.

Georg Wiese, Dirk Weissenborn, and Mariana Neves. 2017a. Neural domain adaptation for biomedical question answering. *arXiv preprint arXiv:1706.03610.*

Georg Wiese, Dirk Weissenborn, and Mariana Neves. 2017b. Neural question answering at bioasq 5b. In *BioNLP 2017*, pages 76–79, Vancouver, Canada,. Association for Computational Linguistics.

Zi Yang, Yue Zhou, and Eric Nyberg. 2016. Learning to answer biomedical questions: Oaqa at bioasq 4b. In *Proceedings of the Fourth BioASQ workshop*, pages 23–37.

Ontology-Based Retrieval & Neural Approaches
for BioASQ Ideal Answer Generation

Ashwin Naresh Kumar,* Harini Kesavamoorthy,* Madhura Das,* Pramati Kalwad,*
Khyathi Raghavi Chandu, Teruko Mitamura and Eric Nyberg
Language Technologies Institute, Carnegie Mellon University
{anareshk,hkesavam,madhurad,pkalwad,kchandu,teruko,ehn}@cs.cmu.edu

Abstract

The ever-increasing magnitude of biomedical information sources makes it difficult and time-consuming for a human researcher to find the most relevant documents and pinpointed answers for a specific question or topic when using only a traditional search engine. Biomedical Question Answering systems automatically identify the most relevant documents and pinpointed answers, given an information need expressed as a natural language question. Generating a non-redundant, human-readable summary that satisfies the information need of a given biomedical question is the focus of the Ideal Answer Generation task, part of the BioASQ challenge. This paper presents a system for ideal answer generation (using ontology-based retrieval and a neural learning-to-rank approach, combined with extractive and abstractive summarization techniques) which achieved the highest ROUGE score of 0.659 on the BioASQ 5b batch 2 test.

1 Introduction

In this paper, we describe our attempts to address the Ideal Answer Generation task of the sixth edition of the BioASQ challenge,[1] which is a large-scale semantic indexing and question answering challenge in the biomedical domain. In particular, the sub-task of Phase B of this annual challenge is to develop a system for *query-oriented summarization*. Traditionally, there are two classes of summarization techniques, each having their own merits and pitfalls: (1) extractive and (2) abstractive. While extractive techniques patch relevant sentences together enabling them to generate grammatically robust summaries, they flounder on maintaining coherence and readability. On the contrary, abstractive techniques extract relevant information from the original text, which is then used to generate a novel natural language summary. While abstractive techniques are more succinct and coherent, automatic text generation is prone to grammatical error. This directly implies that extractive summarization techniques should perform well on automatic evaluation metrics (such as ROUGE), but do less well on human evaluation measures which account for precision, repetition and readability. We explore the hypothesis that a combination of these techniques will provide better overall performance on the ideal answer task, when compared with either approach used in isolation.

The dataset we use for development of the current work is released as a part of the sixth edition of the annual BioASQ challenge (Tsatsaronis et al., 2012). The main categories of answers in this data include *summary*, *factoid*, *list* and *yes/no*. There are a total of 2,251 questions, each of which is accompanied by a list of relevant documents and a list of relevant snippets extracted from each of these documents. Our model is an extension to the highest ROUGE scoring model in the final test batch of the fifth edition of the BioASQ challenge (Chandu et al., 2017), which is based on Maximal Marginal Relevance (MMR) (Carbonell and Goldstein, 1998). In addition, we attempted abstractive techniques that are scoped to improve the readability and coherence aspect of the problem. We made 4 submissions to the challenge.

The paper is organized as follows: Section 2 describes our overall system architecture and the implementation details. Experiments and results are discussed in Section 3 followed by conclusion and future work in 4.

2 System Architecture

The main components of the QA pipeline are outlined in Figure 1. As illustrated, the first step is pre-processing of the question to enrich it with

---

*denotes equal contribution
[1]bioasq.org

Proceedings of the 2018 EMNLP Workshop BioASQ: Large-scale Biomedical Semantic Indexing and Question Answering, pages 79–89
Brussels, Belgium, November 1st, 2018. ©2018 Association for Computational Linguistics

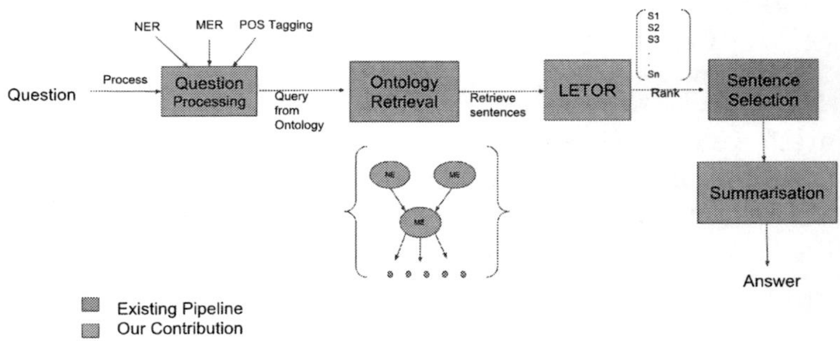

Figure 1: System Architecture

features derived from standard NLP techniques such as Part of Speech (POS) tagging, Named Entity Recognition (NER) and Medical Entity Recognition (MER). Subsequently an ontology-based retrieval system is used to retrieve relevant snippets for the question. The retrieved snippets are combined with the given BioASQ snippets for the question, and passed to the ranking module. The ranked snippets are then input to the sentence selection module from the existing OAQA pipeline (Chandu et al., 2017), which implements the CoreMMR (Zechner, 2002) and SoftMMR algorithms, which use similarity measures to select the most relevant and least redundant snippets. The selected sentences are then passed to the summarization module, which produces the final summary. Each of these modules is discussed in detail below.

2.1 Ontology-Based Information Retrieval

Although a large amount of biomedical text is available in resources such as NLM (NIH, 2018), it can be difficult to leverage in the absence of supervised or automatic labeling (annotation) of the unstructured text content. Our hypothesis is that an Ontology-based retrieval module which utilizes entity and relation extraction techniques to represent and compare the content of questions and candidate answers can improve the recall of answer-bearing documents from unstructured sources.

Our goal is to develop a graphical model that can represent the content of the question and each candidate answer. The nodes in the graph represent medical entities and the edges between them represent the relations between the entities. We extract relations from the text and index them into the graph based on previously-published work (Abacha and Zweigenbaum, 2015). The base architecture for the Ontology-based Retrieval module is shown in Figure 2.

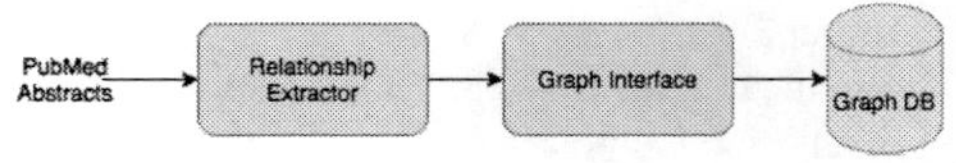

Figure 2: Graph Generation Pipeline

For every edge present in the graph, we store the ID of the source abstract, along with the ordinal index of the source sentence.

2.1.1 Relation Extraction

Relation Extraction (RE) is a technology used by an ontology-based retrieval system to capture the semantic relations which exist between the named entities mentioned in the text; both the entities and relations are considered instances of a given set of ontological types. In practice, various NLP toolkits are available for related tasks, such as dependency parsing, semantic role labeling, and subject-verb-object relation extraction.

However, common NLP tools aren't easily leveraged on biomedical text, due to dramatic differences in the structure and content of the sentences. There exist tools for relation extraction in sub-domains such as Bacteria (Duclos et al., 2007) and disease-cause ontologies (Schriml et al., 2011), but these methods heavily rely on the presence of specific words or features at the sentence level, and cannot be easily scaled to general biotext. Most neural methods for training relation extractors require a large ($O(10^6)$) corpus of labelled examples, which is not available for general biotext (Yih et al., 2015). In order to explore the use of ontology-based retrieval, we developed a novel

RE approach, which is described below. The base architecture for the RE module is depicted in Figure 3.

The following 4 steps are employed for extracting relations from a sentence:

1. Noun Phrase Chunking: The sentence is parsed using the TreeTagger POS tagger (Schmid, 1995) to obtain all the Noun Chunks that form the potential nodes of the graph. For our purpose, the nodes of the graph are all Medical and Named Entities. In order to perform this, the potential nodes are passed through a Medical Entity Recogniser (GRAM-CNN) (Zhu et al.) and the Stanford NER (Manning et al., 2014) discarding the chunks that are not recognized. For an example, let us consider the following sentence: *'Genomic microarrays have been used to assess DNA replication timing in a variety of eukaryotic organisms.'* which extract the following noun chunks: *'Genomic Microarrays'*, *'DNA Replication Timing'* and *'Eukaryotic Organisms'*.

2. Relation Extraction: This step comprises of 2 sub parts.

(2a) RE using Predicate Argument Structures: The Predicate Argument Structure (PAS) for the sentence, obtained using the Enju parser (Miyao et al., 2008), is further parsed in order to obtain possible relations for the graph. Possible relations are those that contain arguments related through a verb or a preposition.

(2b) RE transformation through transitivity: Transitivity is performed on relations obtained from the Enju parser in order to ensure that the arguments of the relations represent medical or named entities in the graph. The potential nodes are passed through the NER and MER. Nodes that are not tagged or recognized by either undergo a transitive transformation to give way to new relations. For the example mentioned, the following relations are formed post transitive formations: *'Microarrays assess Timing'*, *'Timing in Organisms'* and *'Microarrays in Organisms.*

3. Mapping to CUI: As the same medical entity can be represented in many forms, we employ a mapping to the Concept Unique Identifier (CUI) from the UMLS Metathesaurus (Bodenreider, 2004) using the python wrapper for MetaMap called pyMetamap (Aronson, 2001). For each of the Noun Chunks present, the UMLS Metathesaurus is queried to check if a CUI is present. If not, the individual CUIs are obtained for every word forming the noun chunk and the following

rules are employed in order to form a hierarchical node structure. For the Noun chunks obtained in the example, CUIs are directly available for *'DNA Replication Timing'* and *'Eukaryotic Organism'* and not for *'Genomic Microarray'*. To build the tree structure for this node, the CUIs for *'Microarray'* and *'Genomic'* are individually obtained and since the latter is an adjective, the former becomes the child of the latter. The final CUIs obtained and the CUI node tree structure for *'Genomic Microarrays'* are depicted in Table 4 and Figure 6 respectively. For forming relations, the child nodes of all trees are used for connecting edges.

4. Relation Formation: As the arguments in the relations obtained through PAS are the base noun forms that do not represent the whole Noun Chunk, they are expanded to form the whole noun chunk. For the example, the relations obtained in 2b are expanded using the noun chunks to form the final relations as follows: *'Genomic Microarrays assess DNA Replication Timing'*, *'DNA Replication Timing in Eukaryotic Organisms'* and *'Genomic Microarrays in Eukaryotic Organisms'*. The entities in the relation are mapped to their CUI based representation to form a complete relation ready for insertion or retrieval from the graph. The mapped relations for the examples look as follows: *'C1709016, C0887950 assess C1257780'*, *'C1257780 in C0684063'* and *'C1709016, C0887950 in C0684063'*. Here, the root and children nodes are comma separated. The final graph structure for the relations is depicted in Figure 4.

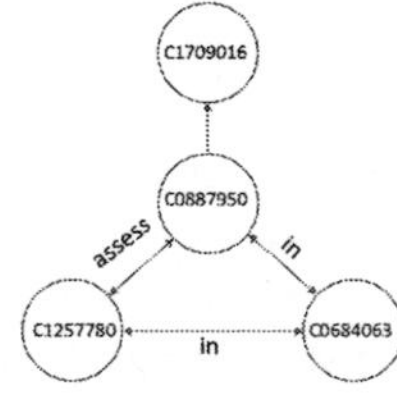

Figure 4: Graph obtained by relation extraction.

In order to index the graph, the relation extraction process specified above is utilized. The edges of the graph store additional information such as the abstract ID and the sentence offset in the abstract. In order to form a relation for a new query to retrieve information from the graph, a back-off mechanism is employed as follows.

1. Form relations using the process for indexing.

2. Query with all medical entities and obtain all

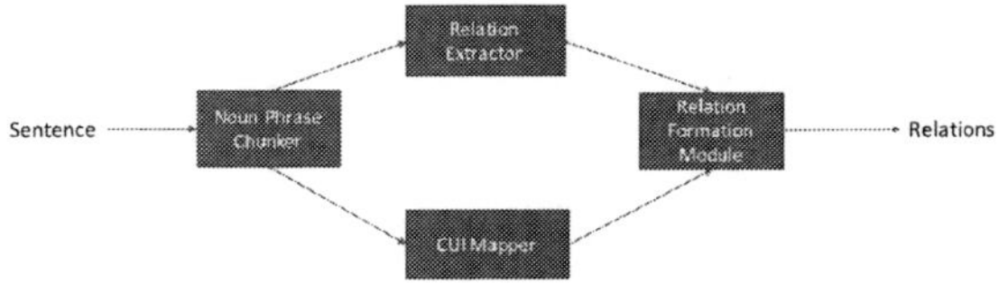

Figure 3: Architecture of Relation Extraction Module

abstracts between pairs of them.

3. Query the medical entities with the verb forms associated with them.

4. Query with just the medical entity and obtain all abstracts related to each of the medical entities.

2.1.2 Graph Creation

Graph Framework: All PubMed abstracts are tokenized and relations are extracted from them. These are added as relations in the graph. We create custom data structures for the Nodes and Edges(Relations) in Neo4j (Webber, 2012). Every relation has attributes which are comma separated values of PubMed ID, location within the abstract. This is stored in order to retrieve the exact sentence that was used to create a particular relation. We hypothesize that this can improve in getting relevant snippets across the abstracts.

Phase I: Ontology Creation with UMLS Concepts. Part of speech tagging is the most intuitive way of approaching the problem of extracting the relations from a given text. An initial strategy of forming the Subject Verb Object (SVO) triplets was formed based on a left-right parsing of the text. For this purpose, an off-the-shelf POS tagger (Schmid, 1994) was used. This is not an effective method to create a graph as it misses very important clauses and fails to recognize the Noun Chunks in sentences. In order to overcome the limitations of this, we form a UMLS based mapping of the noun phrases to medical concept thesaurus as in the UMLS Meta Thesaurus. Once the CUI ids are mapped using metamap as in Section 2.1.1, these are prepared to be indexed into the graph. These relations are added into the graph by following standard database methods, i.e.,

- If the relation is already present in the graph, then the relation attribute is appended with the PubMed ID, and offset in the abstract.

- Every node in the relation is first queried from the graph and then the relation is added between the nodes retrieved. If no node is retrieved, then new nodes are created and the

relation is created.

Phase II: Adding transitivity to Relations. The limitations of Phase I was that despite accurate relations from the relation extractor, the mapping in the graph for different clauses joined together with different prepositions was not solved. We solve it with the following method. First, a key value pair is added as an attribute to the relation, where every key is a preposition and the value is the NodeID noun chunk that is associated with the preposition. For example, in the sentence "Genomic microarrays have been used to assess DNA replication timing in a variety of eukaryotic organisms", the clause "in a variety of eukaryotic organisms" would be missed in the phase II of ontology creation. But in Phase II, we convert such that the verb "assess" has an attribute "$\{in, node_{xyz}\}$" where $node_{xyz}$ is the node pertaining to the CUI of "eukaryotic organisms".

2.2 Ranking

Information Retrieval is one of the essential components of a Question Answering pipeline. It will help provide relevant information to the pipeline for more accurate answers. Ranking snippets based on relevance to the question will improve the answer selection process and in turn give more relevant answers. Employing Learning to Rank (LETOR)(Qin et al., 2010) methods to rank snippets should help rank snippets according to the questions.

The output of the Ontology based Information Retrieval is a set of relevant snippets. We combine the given BioASQ snippets along with the Ontology Retrieved snippets and rank them according to relevance to the question. The ranking algorithm finally gives a set of ranked snippets relevant to the question. There is a possibility that the Ontology based retrieved snippets may also have irrelevant snippets. To prevent the error from further propagating into the pipeline, we use a simple BM25(Robertson and Zaragoza, 2009) scoring threshold between the snippets and the question.

We discard the snippets which have a BM25 score lower than a certain threshold. LETOR is highly feature driven which necessitates a good amount of feature engineering. The explored features are listed in the next section. In this paper, we have explored 2 LETOR approaches: a) RankSVM and b) A Listwise Neural Approach.

2.2.1 Feature Engineering

Multiple features have been explored as inputs to the LETOR framework. The features can be divided into 3 major categories i.e. 1) Statistical Features 2) Semantic Features and 3) Syntactic Features. The statistical features included length of snippets, BM25 score between query and snippet, dot product of tf-idf between query and snippet, cosine similarity over the TF-IDF vectors (along with log space representation), number of bi-grams in the intersection of query and snippets and Jaccard similarity score. The semantic features include averaged word2vec representations across snippets. The syntactic features are number of medical entities and bag of words representation of medical entities.

2.2.2 Quasi Ground Truth Creation

An initial challenge while formulating the LETOR framework is the ground truth ranking of the snippets as that was not provided in the BioASQ training data. The primary purpose of the LETOR model was to rank more relevant snippets higher up to obtain a higher ROUGE score. Taking this into account, we decided to create the ground truth as the BM25 scores between the snippet and the ideal answer. The scores were calculated according to the formula in 1. The snippets were ranked according to the scores for each question.

$$RelevanceScore \\\\ = score_{BM25}(idealanswer, snippet) \tag{1}$$

2.2.3 RankSVM

RankSVM(Cao et al., 2006) is a pairwise LETOR approach towards ranking of documents. Each pair of snippets was taken for a question and was labeled as -1 if the second snippet was ranked higher and +1 if the second snippet was ranked lower. In a pairwise approach there is an overhead of maintaining the metadata as we need to know which set of snippets are going into the SVM as input for validation of the model. Consider $F(Q, S_1)$ as a feature representation of the question Q and snippet S_1. Similarly, $F(Q, S_2)$

is a feature representation of the question Q and snippet S_2. $F(Q, S_1)$ and $F(Q, S_2)$ are inputs to the SVM and the SVM predicts a -1 or +1 according to the relative ranking of S_1 and S_2.

2.2.4 Neural Ranking Approach

The second approach that we implemented was a list-wise ranking approach inspired from List-Net(Cao et al., 2007). Every data point is a feature representation between the question Q and n^{th} snippet S_n. The neural network is trained against the BM25 scores between the snippet and the ideal answer. The architecture of the network is a 2 layer MLP with ReLU activations. The final layer is a linear layer of size 1.

In an ideal scenario, where we would have had the ground truth rankings of the snippets, it would be intuitive to use a probabilistic loss. In our case, as we are using proxy golden ranks with the BM25 score, it would be more intuitive for the model to learn to estimate the scores instead of the relative ranking of the snippets. Hence, we use a RMSE loss as we want our model to estimate the BM25 scores. The RMSE loss is calculated per question as we would want to learn the distribution of snippets with respect to a single question and not across the complete dataset. The final ranking of the snippets are determined with respect to the scores the model predicts.

2.3 Summarization

Summarization is the final stage in the question answering pipeline. The ranked snippet sentences feed into the summarization module which finally outputs the ideal answers.

For the case of ideal answer generation, two types of summarization techniques can be employed; extractive and abstractive summarization. Extractive summarization works by selecting the most relevant sentences in a document to generate the summary (Allahyari et al., 2017). The summaries generated using this technique generally obtain high ROUGE scores (Lin, 2004) due to the high n-gram overlap between the generated summary and the ideal answer. Abstractive summarization on the other hand works by generating the summary word by word as opposed to picking sentences in the case of extractive summarization. Recent advances in abstractive summarization using Pointer Generator Coverage (PGC) networks (See et al., 2017) have shown that neural sequence to sequence models can generate abstractive summaries which are readable and have high ROUGE scores. Given that these networks can generate human readable summaries with high ROUGE scores, we decided to use this model as the ideal answer generation module in our question answering pipeline. We show that the neural sequence model is able to generate ideal answers in the biomedical domain. More concretely, we see that a pretrained PGC model can be transferred to the biomedical domain to generate ideal answers i.e., the model is able to handle new words (mainly biomedical words) and not generate any unknown tags (<UNK>) in the summaries which would hinder the readability. We also show that fine tuning the model on the BioASQ data generates better answers in terms of the ROUGE scores.

3 Experiments and Results

This section describes the experiments conducted to evaluate each of the components in the question answering pipeline. All the experiments have been conducted on batch 2 of the fifth edition of BioASQ and evaluated using the official *Oracle*[2] developed by the organizers of the task.

3.1 Ontology based Information Retrieval

Ontology based retrieval has been evaluated for providing summary answers to queries with zero snippets provided. For example, the snippets retrieved for the question, *'Does metformin interfere thyroxine absorption?'* has a ROUGE of 0.2044 compared with the ideal answer provided for it.

3.2 Ranking

3.2.1 RankSVM Feature Analysis

Ablation studies were carried out with respect to the features to determine which set of features give us the best results.

It is seen that the statistical features contributed the most to the model. Another interesting observation from the graph is that even though BM25 and log(BM25) were the top contributing features, the log(BM25) has a higher weight. This is mainly because the log scale is known to be more stable and therefore will help the model learn better.

3.2.2 Neural Ranking Approach Analysis

The Neural approach was evaluated against the RankSVM results. We also added the syntactic features and did a comparison study on them.

From Table 1 it is seen that adding the syntactic features have contributed to an overall increase in the ROUGE scores for both the models. Also, it is noticed that the Neural model has performed better than the RankSVM. This is mostly due to the fact that the Neural approach is trying to estimate the BM25 scores between the snippet and the ideal answer rather than trying to mimic the quasi ranking. From the discussion of the results above, we can confirm the hypothesis that ranking snippets in an order of relevance will help improve the quality of answers generated by the pipeline.

3.3 Summarization

Ranked snippet sentences from the ranking pipeline are fed into the summarization module. The following experiments were carried out:

1. Using the PGC network pretrained on CNN/Daily Mail to generate the ideal answers

2. Fine tuning the pretrained PGC network on BioASQ data

Table 1 gives the ROUGE scores obtained by both the models on the BioASQ dataset. For the model fine tuning, the pretrained model is further trained on BioASQ 5b training data. We see that the fine tuned model obtains much higher ROUGE-2 and ROUGE-SU4 scores when compared to the pretrained model. This shows that the fine tuned model generates better answers than the pretrained model in terms of ROUGE score. On closely analyzing the answers generated by the PGC models, we see that there are no <UNK>s generated by both the pretrained and fine tuned models. The model is also able to effectively copy the unknown words from the biomedical source text. A detailed error analysis for the answers generated by the model is discussed in the upcoming subsection 3.3.1.

3.3.1 Error Analysis

This subsection discusses the analysis on the ideal answers generated by both the pretrained model and fine tuned model for different question types in the BioASQ dataset (Table 2). All the readability judgments made in this subsection are an indicator of the subjective judgments made by the authors of this paper:

1. **Yes/No type:**

---

[2]BioASQ Oracle

RankSVM LETOR Framework	ROUGE-2	ROUGE-SU4
BM25, Cosine Similarity, TF-IDF dot product	0.6442	0.6453
BM25, Cosine Similarity, TF-IDF dot product, word2vec dot product	0.6439	0.6460
BM25, Cosine Similarity, TF-IDF dot product, word2vec dot product, Snippet Length, log(sum(TF-IDF))	**0.6471**	**0.6481**
Neural LETOR Framework		
Neural Approach	0.6565	0.6552
Neural Approach + MER Bag-of-words	**0.6590**	**0.6581**
Abstractive Summarization		
Pretrained PGC	0.29	0.3
Fine tuned PGC	**0.4640**	**0.4659**

Table 1: Results on 5b batch 2

Question	Ideal Answer	Pre-trained model answer	Fine tuned model answer
Do proton pump inhibitors affect thyroxine absorption? (Yes-No type)	*Proton-pump inhibitors, antacids and a long list of drugs may decrease thyroxine absorption.* Patients with hypothyroidism and normal TSH values during LT4 replacement therapy may need additional thyroid function testing after treatment with PPIs and may need adjustment of their LT4 dose.	*Ppis and a long list of drugs may decrease thyroxine absorption* . Many commonly used drugs , such as bile acid sequestrants , ferrous , phosphate binders , raloxifene and *proton-pump inhibitors , have also been shown to interfere with the absorption of levothyroxine* . Patients with hypothyroidism and normal tsh values during short-term treatment .	*Antacids inhibitors , antacids and a long list of drugs may decrease thyroxine absorption* . Many commonly used drugs , such as bile acid sequestrants , ferrous sulphate , calcium carbonate , aluminium-containing antacids , phosphate binders , raloxifene and *proton-pump inhibitors , have also been shown to interfere with the absorption of levothyroxine* . Pantoprazole did not influence endocrine function in healthy male volunteers during short-term treatment .
What is the ubiquitin proteome? (Factoid Type)	*The ubiquitin proteome is the entire set ubiquitinated proteins and of their respective ubiquitination sites.*	Tandem repeated ubiquitin binding entities tubes under non-denaturing conditions followed by mass spectrometry analysis to study global ubiquitylation events that may lead to the identification of potential drug targets . To study the ubiquitin proteome we have established an immunoaffinity purification method for the proteomic analysis of endogenously ubiquitinated protein complexes .	Mass spectrometry now allows high throughput approaches for the identification of the thousands of ubiquitinated proteins and of their ubiquitination sites . We used tandem repeated ubiquitin binding entities tubes under non-denaturing conditions followed by mass spectrometry analysis to study global ubiquitylation events that may lead to the identification of potential drug targets .
What memory problems are reported in the "Gulf war syndrome"? (List Type)	*Loss of memory and dysmnesia are memory problems* reported in the Gulf war syndrome. Patients suffering from this syndrome often have *other nonspecific symptoms such as fatigue, skin rash, headache, muscle and joint pain and sexual dysfunction.*	The strongest associations were for *mood swings or 20.9*. The symptoms include *fincapacitating fatigue , musculoskeletel and joint pains , headaches , muscle and joint pain* . His history revealed several front-line operations in 1990 and 1991 during the gulf war .	The gulf war syndrome gws is the beginning of what was to be identified as the gulf war syndrome . In early 1992 , u.s. troops returning from the gulf war began reporting a variety of nonspecific symptoms such as *fatigue , skin rash , muscle and joint pain , and loss of memory* .
What is the treatment of acute pericarditis? (Summary Type)	A multidisciplinary approach is frequently necessary to treat acute pericarditis; the most frequent treatments are: *antiinflammatory steroid and non-steroid drugs, antibiotic therapy, pericardial drainage and, less frequently ,intrapericardial irrigation of fibrinolytics; antituberculous chemotherapy in presence of Tuberculous Agent*	The incidence of purulent pericarditis has decreased since the antibiotic era . it is typically an acute and potentially lethal disease , necessitating rapid diagnosis and adequate therapy to improve prognosis . *standard treatment combines appropriate antibiotic therapy with surgical drainage.*	Purulent pericarditis has decreased considerably since the antibiotic era . It is typically an acute and potentially lethal disease , necessitating rapid diagnosis and adequate therapy to improve prognosis . *Standard treatment combines appropriate antibiotic therapy with surgical drainage* .
What causes erucism? (Summary Type)	Erucism is defined as urtication by Lepidoptera larvae. It is a skin reaction to envenomation from certain poisonous caterpillar bristles. The hair on the dorsum of the last instar larvae of the moth may cause urticarial reactions (erucism) as well as eye problems and temporary blindness.	Tropical insects carry infectious diseases , cutaneous injury can occur by other mechanisms . the hair on the dorsum of the moth may cause urticarial reactions . *erucism is defined as urtication by lepidoptera larvae . erucism is defined as urtication by lepidoptera larvae* .	Tropical insects carry infectious diseases , cutaneous injury can occur by other mechanisms . The hair on the dorsum of the last instar larvae of the moth may cause urticarial reactions . *Urticating is defined as urtication by lepidoptera larvae . erucism is defined as urtication by lepidoptera larvae* .

Table 2: Examples of error types observed in the qualitative analysis

Here, we see that the model generated answers address the question and also gives out extra facts not described in the ideal answer, but pertaining to the question. We see that the answer given by the fine tuned model seems more complete than that of the pretrained model as it mentions that antacids (which contain PPI) decrease thyroxine absorption and also that they interfere with a specific type of thyroxine, namely levothyroxine.

2. **Factoid type:**

Here, the generated summaries miss the answer as exact answer is not present even in the snippets. Fine tuned model on the other hand, generates an answer more readable than the pretrained model generated answer.

For most of the other factoid questions, we saw that the question was answered correctly, but the answer describes the facts very differently and also gives different extra facts compared to the ideal answer. Another observation was that presence of several acronyms and abbreviations reduced the ROUGE score.

3. **List type:** Here, we see that the answer generated by the fine tuned model is more readable as there is a seamless flow in the answer where the answer starts off by explaining what the Gulf War Syndrome is and later goes on to list the problems reported in the Gulf War Syndrome. We also notice that the pretrained model misses the symptom 'loss of memory' mentioned in the ideal answer which is however picked up by the fine tuned model.

4. **Summary type:** Here, the generated answers partially answer the question as both

Test Batch	System	ROUGE 2	ROUGE SU4
Batch 1	RankSVM	0.6372	0.6456
Batch 3	Neural Ranking + Extractive Summarization	**0.5743**	**0.5883**
Batch 4	Neural Ranking + Abstractive Summarization	0.4183	0.4281
Batch 5	Relation extraction + Ontology + Neural Ranking + Abstractive Summarization	0.4573	0.4637

Table 3: BioASQ Results for Task 6B

of them mention the surgical drainage treatment which is a super set of the pericardial drainage treatment. Other treatment types are however missed by both the models in their answers. This is mainly due to the fact that there is no direct mapping between the ideal answer and the snippets.

5. **Summary type:** Another error which occasionally surfaces is repetition. In the pre-trained model answer, we see that the last sentence is repeated. In the fine tuned model however, there is no repetition with respect to the entire sentence. Although, a majority of the last sentence is repeated in the fine tuned model answer, we see that the term 'urtication' is defined in terms of its own verb form and urtication is later used to define erucism.

Table 3 comprises our results over the test batches of the sixth edition. We believe the model gave the best ROUGE scores for test batch 3 as we have seen from the previous section that the neural model along with domain specific features performed the best among all the models.

4 Conclusion and Future Work

This paper discusses our system for summary type answer generation using a knowledge graph and a neural learning to rank approach. The ranked snippets are further used to generate the answers using extractive and abstractive summarization techniques. We also show that we can transfer the abstractive summarization knowledge from the CNN/Daily-Mail summarization task to the task of biomedical summarization.

From a brief manual inspection of the generated summaries and their relevant documents, we believe that from an NLP standpoint the following are some of the promising directions to explore. Anaphora resolution would help provide better relations. We also plan to use the ontology indexing and retrieval system for factoid and list types of questions. Incorporating the question type as contextual information while generating summaries could lead to improving precision. Instead

of a dual step of transfer learning with training and fine tuning on PGC network, the abstracts of the PUBMED articles and the entire document could potentially be leveraged to train the end to end network.

As an extension, we intend to pursue the following tasks for BioASQ:

- The current pipeline of the work includes the MMR algorithm while selecting sentences. Experimentation with other diversification algorithms like xQuAD(Santos et al., 2010) and PM-2(Dang and Croft, 2012) can be used for sentence selection.

- Exploration of more language model based features in the LETOR pipeline like Pointwise Mutual Information.

References

Asma Ben Abacha and Pierre Zweigenbaum. 2015. Means: A medical question-answering system combining nlp techniques and semantic web technologies. *Information processing & management*, 51(5):570–594.

Mehdi Allahyari, Seyedamin Pouriyeh, Mehdi Assefi, Saeid Safaei, Elizabeth D Trippe, Juan B Gutierrez, and Krys Kochut. 2017. Text summarization techniques: A brief survey. *arXiv preprint arXiv:1707.02268*.

Alan R Aronson. 2001. Effective mapping of biomedical text to the umls metathesaurus: the metamap program. In *Proceedings of the AMIA Symposium*, page 17. American Medical Informatics Association.

Olivier Bodenreider. 2004. The unified medical language system (umls): integrating biomedical terminology. *Nucleic acids research*, 32(suppl_1):D267–D270.

Yunbo Cao, Jun Xu, Tie-Yan Liu, Hang Li, Yalou Huang, and Hsiao-Wuen Hon. 2006. Adapting ranking svm to document retrieval. In *Proceedings of the 29th annual international ACM SIGIR conference on Research and development in information retrieval*, pages 186–193. ACM.

Zhe Cao, Tao Qin, Tie-Yan Liu, Ming-Feng Tsai, and Hang Li. 2007. Learning to rank: from pairwise approach to listwise approach. In *Proceedings of the 24th international conference on Machine learning*, pages 129–136. ACM.

Jaime Carbonell and Jade Goldstein. 1998. The use of mmr, diversity-based reranking for reordering documents and producing summaries. In *Proceedings of the 21st annual international ACM SIGIR conference on Research and development in information retrieval*, pages 335–336. ACM.

Khyathi Chandu, Aakanksha Naik, Aditya Chandrasekar, Zi Yang, Niloy Gupta, and Eric Nyberg. 2017. Tackling biomedical text summarization: Oaqa at bioasq 5b. *BioNLP 2017*, pages 58–66.

Van Dang and W. Bruce Croft. 2012. Diversity by proportionality: An election-based approach to search result diversification. In *Proceedings of the 35th International ACM SIGIR Conference on Research and Development in Information Retrieval*, SIGIR '12, pages 65–74, New York, NY, USA. ACM.

Catherine Duclos, Jérome Nobécourt, Gian Luigi Cartolano, Anis Ellini, and Alain Venot. 2007. An ontology of bacteria to help physicians to compare antibacterial spectra. In *AMIA Annual Symposium Proceedings*, volume 2007, page 196. American Medical Informatics Association.

Tadayoshi Hara, Yusuke Miyao, and Jun-ichi Tsujii. 2010. Evaluating the impact of re-training a lexical disambiguation model on domain adaptation of an hpsg parser. In *Trends in Parsing Technology*, pages 257–275. Springer.

Chin-Yew Lin. 2004. Rouge: A package for automatic evaluation of summaries. *Text Summarization Branches Out*.

Christopher D. Manning, Mihai Surdeanu, John Bauer, Jenny Finkel, Steven J. Bethard, and David McClosky. 2014. The Stanford CoreNLP natural language processing toolkit. In *Association for Computational Linguistics (ACL) System Demonstrations*, pages 55–60.

Yusuke Miyao, Rune Sætre, Kenji Sagae, Takuya Matsuzaki, and Jun'ichi Tsujii. 2008. Task-oriented evaluation of syntactic parsers and their representations. *Proceedings of ACL-08: HLT*, pages 46–54.

NIH. 2018. *National Library of Medicine*.

Tao Qin, Tie-Yan Liu, Jun Xu, and Hang Li. 2010. Letor: A benchmark collection for research on learning to rank for information retrieval. *Information Retrieval*, 13(4):346–374.

Stephen Robertson and Hugo Zaragoza. 2009. The probabilistic relevance framework: Bm25 and beyond. *Found. Trends Inf. Retr.*, 3(4):333–389.

Rodrygo LT Santos, Jie Peng, Craig Macdonald, and Iadh Ounis. 2010. Explicit search result diversification through sub-queries. In *European Conference on Information Retrieval*, pages 87–99. Springer.

Helmut Schmid. 1994. Probabilistic pos tagging using decision trees. In *Proceedings of International Conference on New methods in Language Processing*.

Helmut Schmid. 1995. Treetagger| a language independent part-of-speech tagger. *Institut für Maschinelle Sprachverarbeitung, Universität Stuttgart*, 43:28.

Lynn Marie Schriml, Cesar Arze, Suvarna Nadendla, Yu-Wei Wayne Chang, Mark Mazaitis, Victor Felix, Gang Feng, and Warren Alden Kibbe. 2011. Disease ontology: a backbone for disease semantic integration. *Nucleic acids research*, 40(D1):D940–D946.

Abigail See, Peter J Liu, and Christopher D Manning. 2017. Get to the point: Summarization with pointer-generator networks. *arXiv preprint arXiv:1704.04368*.

George Tsatsaronis, Michael Schroeder, Georgios Paliouras, Yannis Almirantis, Ion Androutsopoulos, Eric Gaussier, Patrick Gallinari, Thierry Artieres, Michael R Alvers, Matthias Zschunke, et al. 2012. Bioasq: A challenge on large-scale biomedical semantic indexing and question answering. In *AAAI fall symposium: Information retrieval and knowledge discovery in biomedical text*.

Jim Webber. 2012. A programmatic introduction to neo4j. In *Proceedings of the 3rd annual conference on Systems, programming, and applications: software for humanity*, pages 217–218. ACM.

Scott Wen-tau Yih, Ming-Wei Chang, Xiaodong He, and Jianfeng Gao. 2015. Semantic parsing via staged query graph generation: Question answering with knowledge base.

Klaus Zechner. 2002. Automatic summarization of open-domain multiparty dialogues in diverse genres. *Computational Linguistics*, 28(4):447–485.

Qile Zhu, Xiaolin Li, and Ana Conesa. Gram-cnn: a deep learning approach with local context for named entity recognition in biomedical text. *Bioinformatics*, page btx815.

Supplementary Material

For the example provided in section 2.1.1, the Predicate Argument Structure provided by Enju is shown in Figure 5, which forms the following relations: *'Microarrays assess Timing'*, *'Assess in variety'* and *'variety of organisms'*.

Table 4: CUI Mapping

Concept	CUI
Microarray	C1709016
Genomic	C0887950
DNA Replication Timing	C1257780
Eukaryotic Organism	C0684063

Figure 6 shows the node structure that is built from the CUIs for Genomic Microarray.

Figure 6: CUI node structure for Genomic Microarray

The efficacy of the Relation Extraction module depends on the tools it utilizes. The Enju parser trained using the GENIA corpus has an F-score of 90.15 on the same (Hara et al., 2010). GRAM-CNN has an F1-score of 87.26% on the Biocreative II dataset, 87.26% on the NCBI dataset and 72.57% on the JNLPBA dataset (Zhu et al.). In addition to the hierarchical node structure, these can also lead to the existence of incorrect nodes and edges in the ontology. Every node is associated with all abstracts containing a mention of them. This results in the possibility of the retrieval logic returning all abstracts containing just a mention of the medical/named entity present in a query, rather than only the relevant abstracts for that particular query. The Ranking module filters these abstracts to obtain those most relevant to the query.

We also graphed out the top contributing features for our RankSVM model. Figure 7 displays the contribution of the top 6 features which were used in the model. The factors by which the top 6 features contributed were then normalized and plotted. The top contributing features are depicted in Figure 7.

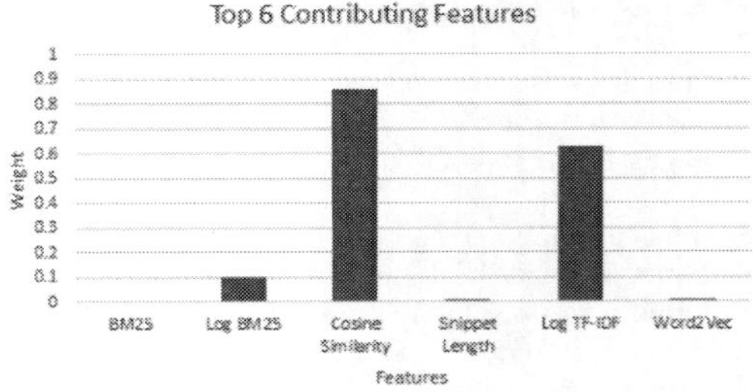

Figure 7: RankSVM Top Contributing Features

ROOT	ROOT2	ROOT3	ROOT4	-1	ROOT5	ROOT6	used	use	VBN	VB	4
used	use	VBN	VB	4	verb_arg123	ARG1	UNKNOWN	UNKNOWN	UNKNOWN	UNKNOWN	-1
used	use	VBN	VB	4	verb_arg123	ARG2	microarrays	microarray	NNS	NN	1
used	use	VBN	VB	4	verb_arg123	ARG3	assess	assess	VB	VB	6
assess	assess	VB	VB	6	verb_arg12	ARG1	microarrays	microarray	NNS	NN	1
assess	assess	VB	VB	6	verb_arg12	ARG2	timing	timing	NN	NN	9
a	a	DT	DT	11	det_arg1	ARG1	variety	variety	NN	NN	12
DNA	dna	NN	NN	7	noun_arg1	ARG1	timing	timing	NN	NN	9
Genomic	genomic	JJ	JJ	0	adj_arg1	ARG1	microarrays	microarray	NNS	NN	1
replication	replication	NN	NN	8	noun_arg1	ARG1	timing	timing	NN	NN	9
eukaryotic	eukaryotic	JJ	JJ	14	adj_arg1	ARG1	organisms	organism	NNS	NN	15
to	to	TO	TO	5	comp_arg1	ARG1	assess	assess	VB	VB	6
of	of	IN	IN	13	prep_arg12	ARG1	variety	variety	NN	NN	12
of	of	IN	IN	13	prep_arg12	ARG2	organisms	organism	NNS	NN	15
in	in	IN	IN	10	prep_arg12	ARG1	assess	assess	VB	VB	6
in	in	IN	IN	10	prep_arg12	ARG2	variety	variety	NN	NN	12
been	be	VBN	VB	3	aux_arg12	ARG1	microarrays	microarray	NNS	NN	1
been	be	VBN	VB	3	aux_arg12	ARG2	used	use	VBN	VB	4
have	have	VBP	VB	2	aux_arg12	ARG1	microarrays	microarray	NNS	NN	1
have	have	VBP	VB	2	aux_arg12	ARG2	used	use	VBN	VB	4

Figure 5: Predicate Argument Structure

Author Index